Hierarchical Task Analysis

Hierarchical Task Analysis

Andrew Shepherd

CRC Press

Taylor & Francis Group
Boca Raton London New York

CRC Press is an imprint of the
Taylor & Francis Group, an **informa** business

First published 2001
by Taylor & Francis
11 New Fetter Lane, London EC4P 4EE

Simultaneously published in the USA and Canada
by Taylor & Francis Inc,
29 West 35th Street, New York, NY 10001

Taylor & Francis is an imprint of the Taylor & Francis Group

TJ International Ltd, Padstow, Cornwall

Publisher's Note

This book has been prepared from camera-ready copy provided by the author.

British Library Cataloguing in Publication Data
A catalogue record for this book is available from the British Library

Library of Congress Cataloging in Publication Data

Shepherd, Andrew, 1946-

  Hierarchical task analysis / Andrew Shepherd.
    p. cm.
  1. Human Engineering. 2. Task analysis. 3 Cognitive science. I. Title
  TA166 .S45 2000
  620.8'2—dc21   0-033766

ISBN 0-7484-0837-1 (hbk)
ISBN 0-7484-0838-X (pbk)

Contents

Figures   ix

Tables   xv

Preface   xvii

Acknowledgements   xix

Introduction   1
    Organisation of the book   3

Chapter 1   Task analysis, concepts and terminology   7
    Introduction   8
    Systems thinking   8
    Systems thinking in describing human skill   11
    Human performance in systems   13
    Justification of HTA in systems terms   16
    Some terminology so far   21
    Concluding comments   23

Chapter 2   HTA - a task analysis framework   25
    Introduction   26
    The main strategies of HTA   26
    HTA - a Framework for Analysing Tasks   32
    Using the framework   39
    Concluding comments   39

Chapter 3   Plans and complexity   41
    Introduction   42
    Different sorts of plan   42
    Composite plans   56
    Unravelling complex plans   56
    Concluding comment   64

Chapter 4   Flexibility, constraint, cognition and context 67
    Introduction   68
    Inferring cognitive operations   72
    Focus and Bias   74
    Modelling and evaluating strategies   74
    Situating cognition   77
    Identifying strategies off line   85
    Concluding remarks   85

Chapter 5   Representing and recording HTA   **87**
    Introduction   88
    Reasons for representing and recording   88
    Hierarchical diagrams   89
    Numbering the analysis   92
    Tabular Formats   93
    Representation of plans in diagrams and tables   97
    Computer aids in recording task analysis   98
    Concluding remarks   100

Chapter 6   Analysis of tasks - some illustrations   **101**
    Introduction   102
    Changing a cartridge   104
    Process control   104
    Task analysis of a batch operation   106
    Task analysis of a continuous process control task   109
    Air-traffic control   111
    Minimal Access Surgery (MAS)   113
    A customer service task   116
    Using a wordprocessor   117
    Mechanical maintenance   120
    Nursing   124
    Management   124
    Staff supervision - nurse in charge of a ward   131
    Concluding remarks   133

Chapter 7   Making human factors design decisions within HTA   **135**
    Introduction   136
    Considering the design options   137
    Making design choices   143
    Context and constraint and design decisions   147
    Collating and resolving design decisions   151
    Developing detailed design within the system life-cycle   152
    Concluding remarks   155

Chapter 8   Teams and jobs   **157**
    Introduction   158
    Collaborating on a common goal   159
    Assessing workload   164
    The tasks of within a working team   168
    Concluding remarks   173

Chapter 9   Information and skill   **175**
    Introduction   176
    The reliance on information in tasks - some examples   177
    Characteristic problems with information and control   179
    Identifying information requirements in operations   185
    Types of operation   187
    Identifying information requirements in plans   193
    Representing and situating information and control   193
    Concluding remarks   195

Chapter 10   HTA and training   **197**
    Introduction   198
    Learning practical skills   198
    An informal training intervention   200
    Elements of a formal training programme   205
    Part-task training   209
    Simulation for training   213
    Concluding comment   216

Chapter 11   Designing support documentation   **217**
    Introduction   218
    Common types of support documentation   219
    Determining where job-aids should be employed   225
    Relating job-aids to training   226
    Representation of job-aids and the link to HTA   228
    Concluding remarks   230

Chapter 12   Human resource management issues   **231**
    Introduction   232
    Establishing a proper task description   232
    Making human resource management decisions   234
    Concluding comment   238

Chapter 13   Conclusions   **239**
    The method   239
    The application of HTA to different domains   241
    The application of HTA to supporting different design solutions   242
    Cognition and flexibility   243
    Tasks and contexts   243
    HTA and other task analysis methods   244

Chapter 14 Notes   **245**

    Chapter 1: Task analysis, concepts and terminology   245
    Chapter 2: HTA - a task analysis framework   247
    Chapter 3: Plans and complexity   254
    Chapter 4: Flexibility, constraint, cognition and context   254
    Chapter 5: Representing and recording HTA   255
    Chapter 6: Analysis of tasks - some illustrations   256
    Chapter 7: Making human factors design decisions within HTA   257
    Chapter 8: Teams and jobs   257
    Chapter 9: Information and skill   258
    Chapter 10: HTA and training   258
    Chapter 11: Designing support documentation   259
    Chapter 12: Some human resource management issues   259

References   **260**

Index   **267**

Figures

Figure 0.1   *A simple illustration of Hierarchical Task Analysis.*   *2*

Figure 1.1   *Systems and subsystems in a hypothetical transportation system.*   *8*

Figure 1.2   *Inputs and outputs to related subsystems.*   *9*

Figure 1.3   *The expanded operator-system interaction, showing information and feedback loops.*   *13*

Figure 1.4   *The human-task system expanded to show additional elements of human performance.*   *14*

Figure 1.5   *The human-task system showing the influences of the environment.*   *16*

Figure 1.6   *The hypothesis and test cycle*   *19*

Figure 2.1   *The basic cycle of decisions during task analysis.*   *26*

Figure 2.2   *A simple information-model of an operation.*   *29*

Figure 2.3   *The cycle of task analysis decisions. .*   *33*

Figure 2.4   *Higher levels of redescription of a maintenance task carried out by packaging line engineers.*   *35*

Figure 2.5   *The packaging line operator's task.*   *35*

Figure 2.6   *How the line operator initially dealt with faults on the line.*   *36*

Figure 2.7   *How the line operator should deal with faults on the line to provide the engineer with better support.*   *36*

Figure 3.1   *Fixed sequence element in a plan.*   *43*

Figure 3.2   *A contingent fixed sequence element.*   *44*

Figure 3.3   *Typical contingent sequence plan elements.*   *45*

Figure 3.4   *This represents the same activities as Figure 3.2 but plan is represented as a fixed sequence.*   *46*

Figure 3.5   *A supermarket checkout task.*   *47*

Figure 3.6   Part of the task analysis of developing a black and white film..   48

Figure 3.7   A time-line plan for starting up 3 interdependent units.   50

Figure 3.8   Controlling a distillation train.   51

Figure 3.9   Deal with a tele-sales customer.   52

Figure 3.10 Typically, procedural cycles fit into jobs as a component of a general shift cycle.   53

Figure 3.11 The common remedial cycle.   53

Figure 3.12 The common remedial cycle in a task analysis.   54

Figure 3.13 Maintaining a baby's temperature in hospital — an illustration of a remedial task in which the cycle of activity is implicit within the plan.   54

Figure 3.14 The chlorine production complex.   57

Figure 3.15 A first attempt at redescription of the chlorine balancing task.   58

Figure 3.16 An interim revision of the chlorine balancing task analysis.   59

Figure 3.17 Further revision of the top level of the analysis of the chlorine balancing task.   60

Figure 3.18 Re-examination of 'Establish new resource balance for system.   61

Figure 3.19 The revised task analysis for the chlorine balancing task.   62

Figure 3.20 Summary of plan elements.   65

Figure 4.1   Representation of a conveyor belt operation in an automated warehouse.   75

Figure 4.2   One strategy for identifying spaces for inserting dockets.   76

Figure 4.3   A second strategy for locating the correct gap between orders.   77

Figure 4.4   Extracts from the HTA of ultra-sonic testing of welds in the nuclear industry.   78

Figure 4.5   Equipment used in the ultra-sonic testing task.   79

Figure 4.6   HTA of an underground railway control task.   82

Figure 4.7   The general functions of a neonatal intensive care team.   84

Figure 5.1   Three versions of the same hierarchy showing different layout compromises.   90

Figure 5.2   Representation of HTA over several pages.   91

Figure 5.3   This shows both the manner in which goals not further redescribed are underlined and illustrates the numbering system.   92

Figure 5.4   The numerical representation of the supermarket checkout task from Figure 3.5 translated into a sequence for an HTA table.   94

Figure 5.5   This compares a flow diagram with a rule set. This example is from the HTA described in Figure 3.5.   98

Figure 5.6   The outline of the supermarket task.   99

Figure 6.1   Changing a cartridge in a laser printer.   104

Figure 6.2   A typical arrangement of processing units in a process plant.   105

Figure 6.3   The top level of an HTA for a batch plant operation.   107

Figure 6.4   The manufacturing phase of the batch plant task analysis.   107

Figure 6.5   This extent of the HTA developed for the batch processing task.   108

Figure 6.6   Extracts from an HTA of a typical continuous process control task.   109

Figure 6.7   Redescription of dealing with off-specification conditions in a continuous process control task.   110

Figure 6.8   The air-traffic control task.   112

Figure 6.9   The minimal access surgery task.   114

Figure 6.10   Detail from the minimal access surgery task dealing with a number of psychomotor skills.   115

Figure 6.11   A telephone customer service task.   116

Figure 6.12   Preparing and printing a letter in a wordprocessor.   118

Figure 6.13   Modifying the order of words in the text.   118

Figure 6.14   Ways of developing text formatting.   119

Figure 6.15   Installing plant and pipework — illustration of a typical mechanical maintenance task.   120

Figure 6.16   Representation of the general maintenance job   122

Figure 6.17   Analysis of 'overhaul valves in workshop which focuses on maintenance of a specific type of equipment in a particular domain.   123

Figure 6.18 *Carrying out nursing duties in a hospital ward.*   *125*

Figure 6.19 *The overall functions of the management task.*   *126*

Figure 6.20 *Setting up and modifying a department.*   *127*

Figure 6.21 *Activities in monitoring and maintaining the effectiveness of the
              department.*   *129*

Figure 6.22 *Developing the department's systems.*   *130*

Figure 6.23 *HTA of nurse in charge of a ward — an example of staff
              supervision.*   *131*

Figure 6.24 *Running the shift.*   *131*

Figure 6.25 *Liaising with other shifts.*   *132*

Figure 7.1 *The interaction between task factors.*   *138*

Figure 7.2 *Strategies for managing the operator's experience.*   *138*

Figure 7.3 *Layout of the batch plant.*   *144*

Figure 7.4 *Redescription of 'reflux'.*   *144*

Figure 7.5 *HTA within an iterative human factors/human resource design
              cycle.*   *155*

Figure 8.1 *A combined analysis of operating and maintaining the
              packaging line.*   *159*

Figure 8.2 *The resultant task for the line operator.*   *160*

Figure 8.3 *The resultant task for the maintenance technician.*   *160*

Figure 8.4 *A proposed supervisory function for an automatic railway.*   *162*

Figure 8.5 *The principle of delegation.*   *163*

Figure 8.6 *A time-line showing how the analyst and task experts judge the
              coincidence of tasks in a railway control task.*   *166*

Figure 8.7 *Timeline recording actual response of controllers.*   *167*

Figure 8.8 *Delegating to a colleague.*   *169*

Figure 8.9 *Delegating to a trainee.*   *169*

Figure 8.10 *Elements of shift handover.*   *171*

Figure 8.11 *Coaching and instructing.*   *172*

Figure 9.1   *A hierarchical menu structure.*   *181*

Figure 9.2   *Flows of information, energy and product through plants.*   *182*

Figure 9.3   *The arrangement of intermediate product stores in the system.*   *183*

Figure 9.4   *Layout of control panels*   *184*

Figure 9.5   *Layout of computer control consoles.*   *184*

Figure 9.6   *HTA of a tele-sales task in a call centre.*   *194*

Figure 9.7   *A possible way of organising menus for the tele-sales task in Figure 9.6.*   *196*

Figure 10.1 *An HTA of learning by experience.*   *199*

Figure 10.2 *HTA of a shop assistant's task.*   *201*

Figure 10.3 *Skills and knowledge mapped onto the task analysis.*   *203*

Figure 10.4 *Interaction between the instructional cycle and the learning process.*   *206*

Figure 10.5 *Top level in the HTA of managing trains in a light rail depot.*   *210*

Figure 10.6 *Monitoring operations within the depot.*   *210*

Figure 10.7 *Some part-task training arrangements.*   *212*

Figure 11.1 *Preparing and developing a black and white film.*   *220*

Figure 11.2 *Two job-aids derived from the HTA in Figure 11.1.*   *220*

Figure 11.3 *A checklist derived from HTA.*   *221*

Figure 11.4 *A decision flow-chart derived from the plan in Figure 3.18.*   *222*

Figure 11.5 *The hierarchical aspects of a task description used to aid the organisation of a manual.*   *224*

Figure 11.6 *HTA of manufacturing a specific product in a batch plant.*   *227*

Figure 11.7 *A batch manufacturing instruction for a single product.*   *228*

Figure 11.8 *A generic HTA for dealing with all products on the batch plant.*   *229*

Figure 12.1 *Outline job-description for a railway supervisory task.*   *237*

Tables

Table 3.1   Clusterings of operations in the chlorine balancing task.   59

Table 3.2   Event table for the chlorine balancing task.   63

Table 5.1   The main columns of the task analysis table.   95

Table 5.2   A simple tabular format showing the use of a 'notes' column. describing a range of human factors issues in a potentially hazardous environment.   96

Table 5.3   Using columns in a task analysis table to allow the analyst to systematically reord comments about operations according to a given set of classifications.   96

Table 6.1   Table of sub-tasks to be analysed for the mechanical maintenance project.   122

Table 7.1   Factors affecting performance and some consequences for their combination.   150

Table 7.2   An example of an HTA table recording the factors affecting performance as an aid to developing human factors hypotheses.   151

Table 7.3   Types of solution to support task performance and their relation to the system design cycle.   154

Table 9.1   Typical information requirements for different operational types.   191

Table 9.2   Description of information types.   191

Table 10.1   The HTA table for the shopping task in Figure 10.2.   202

Table 10.2   Typical stages in a training design and development process.   207

Table 10.3   Summary of the main benefits of HTA in training design.   209

Table 12.1   Selection considerations in a health physics task.   236

Preface

Hierarchical Task Analysis (HTA) was initially developed by Keith Duncan and John Annett in the late 1960s. Their approach incorporated several important ideas from other task analysis methods into a practical analytical framework. The approach has been described in a number of useful articles, but these have often lacked sufficient explanation and illustration to enable people to understand its full implications. One consequence is that practitioners often see HTA as just another tool in the tool-box whose main purpose is to present hierarchical diagrams. Producing diagrams never was the purpose of HTA. The original authors' intentions was always to provide a rigorous and realistic method of examining practical tasks. The diagramatic representation was a by-product.

My aim in preparing this book has been to provide a reasonably comprehensive account of HTA and to show how it can be used more widely and more systematically. Task analysis methods should aim to help the analyst engage with a problem in order to identify a solution. The analyst's purpose is essentially practical. Therefore, I have represented HTA as a *framework* for examining tasks in which different considerations and other analytical methods can be applied as the task analysis evolves. A particular aspect of this approach concerns how well the analyst understands the context in which he or she is working.

I have also tried to show the breadth and power of the method by explaining how different aspects of HTA combine to create a rich picture of a task in order to account for complex behaviours. To support this, I have provided illustrations from a wide range of work contexts - manufacturing, maintenance, medical contexts, transportation and commercial activities. All of the examples included are reasonably straightforward. Where I have felt that some readers may be unfamiliar with certain aspects of particular domains, I have included a simple explanation. I believe there is benefit in exploring tasks across domains - even substantially different domains - since lessons learned in one domain often apply to others.

Finally, I have included several chapters concerned with applying HTA to support different aspects of human factors design - job and team design, interface design training, job-aids and human resource management issues. These are not comprehensive accounts of these topics and only deal with them in outline to show how HTA can be applied. Ultimately, task analysis methods can only be judged with respect to the use that can be made of results.

To avoid cluttering the main text with excessive dates and names I have included a final chapter which contains a number of notes that clarify certain numbered points throughout the text. Readers may pursue or avoid these as they choose. References are confined to these notes.

In view of focus and limitations of space, I have limited my discussion of other task analysis methods and only provide a cursory account of the different areas of applications. To compensate for this, I have referred heavily to texts which provide the reader with extensive supplementary material. These are Kirwan and Ainsworth (1992), Salvendy (1997) and Wilson and Corlett (1995).

References

Kirwan, B. and Ainsworth, L. K. (1992)*The Task Analysis Guide*. London: Taylor and Francis.

Salvendy, G. *(*1997) *Handbook of Human Factors and Ergonomics*, (2nd ed). New York: John Wiley & Sons.

Wilson, J. R., and Corlett, E. N. (1995) *Evaluation of Human Work*. London: Taylor and Francis.

Acknowledgements

I would like to acknowledge the contribution made by both John Annett and Keith Duncan to the theoretical development and the practical conduct of task analysis over the past few years. I have been fortunate to work with both Keith and John on various occasions. Despite offering some variation from how they presented their original ideas, I feel that the account I have provided in this book is entirely consistent with their intentions. I prepared a list of many other friends and colleagues in the human factors community with whom I have worked over the years, but the list got too large. I trust they will see their influence in various parts of the book and I thank them for their support and for providing me with the opportunity to work on some very interesting projects. Finally, I should like to thank the numerous operators, supervisors, managers, nurses, doctors and so on, who have patiently described their responsibilities and given me the opportunity to observe them at work.

Introduction

Hierarchical Task Analysis (HTA) was first proposed in the late 1960s as a general approach to examining tasks. Since then, it has become widely adopted although the method is often applied unsystematically or in ways which fail to ensure its full benefit. The aim of this book is to present the ideas of HTA more fully than previously with a view to demonstrating the method, to show how it can be applied to a number of different work contexts and to show how it can be applied in a number of useful ways. HTA cannot deal with every human factors decision without reference to other methods and ideas, but it can be used to guide an examination of tasks so that other methods and ideas can be used to greater benefit.

Any effort to improve human performance in a work or recreational setting must start by some understanding of what people are required to do and how they achieve their goals. Methods for achieving this understanding are often referred to as *task analysis*. Thus task analysis methods are an important prerequisite to the organisation of work, the design of workplaces, work practices and equipment, and in helping people to master their tasks. Task analysis methods, therefore, should be of direct interest to managers and engineers concerned with setting up and organising systems, to designers concerned with making sure people can use equipment properly, to managers and supervisors concerned with making sure that systems work according to design, to human factors and other management support staff concerned with prescribing conditions to enable people to work effectively, to human resources staff concerned with personnel and training issues, and to safety staff concerned to ensure that safe practices are followed.

In HTA, tasks are represented in terms of hierarchies of *goals* and *subgoals*, using the idea of *plans* to show when subgoals need to be carried out. Figure 0.1 shows these elements representing the simple task of using a toaster. Using HTA to examine something as straightforward as this is not something we would normally bother to do - although using a toaster could be seen as more complex than this - but it provides a good example.

In task analysis, it is always important to think of the reason why the task is carried out. A toaster is used to obtain toast, by cooking ordinary bread to the satisfaction of the person who is to eat it. Thus, the task has a *purpose* or *goal* and *criteria* against which the toast can be judged to be satisfactory or otherwise. Setting the criteria for industrial, commercial and service goals includes specification of the product and constraints on how it is achieved. These constraints can include cost and safety criteria.

1

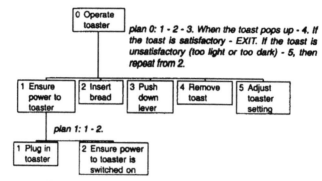

Figure 0.1   A simple illustration of Hierarchical Task Analysis.

Thus, motor cars are manufactured to be capable of transporting passengers according to a criterion of speed and acceleration, but this cannot be achieved at the expense of comfort and safety.

Detailed criteria can rarely be specified at the outset of a design process, even in product design. As designs are developed and intermediate design problems are solved, so new aspects of the product and its manufacture are discovered. To achieve a suitable level of power for a new vehicle, for example, a larger engine than had been initially envisaged may need to be included. This immediately places greater constraints on the size and layout of other components, so detailed design criteria are modified.

This process of refining criteria also arises when tasks are examined. As aspects of the task are uncovered, so we realise increasingly what needs to be valued in terms of performance. For example, a task analysis might commence with the aim of improving human performance to gain greater productivity. Notions of safety may be uppermost, but only when task detail is understood are the implications of safety properly appreciated.

Just as a task has a purpose, so too does the task analyst's intervention in doing task analysis. The analyst might be involved in training, or developing a better control panel, or determining how people can work together most effectively, or several of these things. Task analysis should not be done for the sake of it; knowing why we are carrying out the analysis affects how the analysis progresses.

Plans are crucial to HTA. A plan only makes sense in conjunction with the subgoals it is governing. Thus, to operate the toaster, we can use a plan (plan 0 in Figure 0.1) which states that first we must ensure power to the toaster, then we must insert the bread, then we push down the lever, then when the toast pops up, we remove the toast. If the toast is satisfactory we can terminate the toaster operation. If the toast is unsatisfactory we can adjust the toaster then repeat part of the previous activity.

Carrying out HTA on any task entails similar processes to those described for using the toaster. HTA works towards understanding what is necessary to achieve the stated goal. The analyst keeps in mind the performance criteria involved. As the analysis proceeds, the criteria for performance and why these different things are important start to make more sense.

Organisation of the book

Following this introductory chapter, the book is organised into two main sections. Chapters 2 to 6 deal with the issues of how to analyse tasks, while Chapters 7 to 12 deal with issues of applying the analysis. Since this book aims to provide a practical account of HTA, academic references and discussion of a number of issues have been kept to a minimum to allow the reader to follow arguments and examples more easily. To balance this, Chapter 14 has been included to deal with various numbered annotations throughout the book.

Chapter 1 introduces several of the main concepts used in task analysis to show how the basic methods of HTA are justified. Emphasis is placed on the importance of *systems thinking,* because task analysis is always undertaken to help understand human performance in a context such as an office, hospital ward or industrial unit. Carrying out task analysis in a practical way that serves the needs of people responsible for systems working effectively, requires a broad perspective. Chapter 1 also introduces most of the terminology that will be used throughout the book.

Task analysis methods have both *products* by which the task is represented and *processes* by which the analyst gathers information about the task in order to develop this representation. HTA is strongly characterised by hierarchical diagrams, such as that shown in Figure 0.1. This has led people to assume that HTA is, primarily, a method for task representation. However, *Chapter 2* will argue that this hierarchical representation is merely the product of a systematic process of examining tasks. Indeed, it will suggest that HTA is most usefully regarded as a framework for guiding task analysis. This framework enables tasks to be explored in accordance with their degree of complexity.

Chapter 3 focuses on the issue of *plans* in HTA. Plans are crucial in redescribing goals, because they specify the conditions when subgoals must be carried out to meet their common goal. It will be shown that much of the flexibility and variety between different tasks can be accounted for by different plans and that plans can combine to account for complex behaviours. Strategies for examining complex plans will also be set out.

Trying to be a clear as possible when describing tasks risks plans appearing too inflexible. As a consequence, task analysis may fail to represent the flexibility that experienced workers have to bring to their work. Real skills often reflects considerable planning and decision-making which causes observable performance to vary in accordance with the context in which the task is carried out. *Chapter 4* will deal with this issue. HTA is not a method of *cognitive task analysis* but serves the cause of cognitive task analysis by setting out the context in which such considerations apply. Moreover, attempts at representing how people reason and plan are often unnecessary in practical projects. The manner in which HTA relates to the analysis of cognitive tasks will be discussed.

Chapter 5 discusses issues of recording and representing HTA. Maintaining good records of work in progress and work completed is good practice and essential to task analysis where different aspects of how the task is carried out need to be negotiated with different people in authority and with the necessary expertise. Maintaining good records is essential to record how decisions are made. Good recording systems are also important because task descriptions can become complex; methods of locating parts of the task must be easy to follow without losing essential task detail.

Chapter 6 provides the reader with several examples of task analysis carried out in different work contexts. The aims of Chapter 6 are to demonstrate breadth of application of HTA and to show how similar analyses can be developed elsewhere. Examples are taken from clerical, industrial, health, transportation and service contexts. Operational, maintenance, management and supervisory tasks are included.

The second main section considers application of HTA. Often accounts of task analysis are given simply in terms of gathering and representing information about the task with no guidance provided to show how this information can then be used. This is generally unsatisfactory, because the methods adopted and the decisions taken during a task analysis are closely interrelated with the purpose for which the analysis is being carried out. Moreover, task analysis is a practical activity for which utility is of prime concern. If task analysis methods cannot be used to help identify performance problems or improve the design of tasks, then their value is questionable, no matter how valid the task analysis appears to be. Therefore, Chapters 7 to 12 will deal with various aspects of application of HTA.

Chapter 7 will consider the issue of *human factors design* and how design choices are influenced by various design factors. Design decisions are usually taken as HTA progresses; the HTA framework set out in Chapter 2 shows that making design choices is an integral part of the decision-making in which all analysts must engage. Chapter 7, will develop further how these choices are made, by reference to a range of contextual factors that influence both how tasks are performed and the costs and benefits of the design solutions themselves. Examples are given to illustrate how these factors interact.

Chapter 8 considers the issue of *team and job design*. Task analysis traditionally focuses on the role of the individual operator. However, by using a *functional* approach such as HTA, it is possible to appreciate the demands placed on an individual working within wider team whose role is to service a common function within an organisation. Examples are given of how team functions may be devolved into individual functions. It will also be shown how the various means by which colleagues are able to work effectively with one another can be represented as 'team tasks' in order that they are understood and can be trained. By understanding individual jobs in this way, their contribution to the wider team is clarified.

Chapter 9 considers issues associated with *work design*, including the issue of interface design. As with other design decisions, these aspects of human factors design are considered as an HTA progresses. Of particular importance is the fact that human skill depends upon the information and resources available to the operator carrying out a task. Indeed, many workplace design issues relate to task analysis in terms of the extent to which they support or constrain the operator's use of *information and controls*. HTA can make a useful contribution to this aspect of design by helping to identify the information upon which the operator must rely.

Chapter 10 considers issues of *training*. Training has always been a central focus for HTA. This chapter will illustrate how HTA can be used to identify training needs, identify task knowledge, specify conditions for practice, including helping to prescribe simulation, and assist with assessment.

Chapter 11 considers the development of *job-aids* and other user documents. By providing a clear statement of the task, HTA can be used to identify where job-aids will assist performance. Moreover, the structure of HTA offers substantial benefit in helping set out a job-aid consistent with task requirements and with training.

Chapter 12 deals with the issue of *personnel* or *human resource management*. HTA provides a representation of the task that can aid decisions concerned with aspects of recruiting and managing people to ensure they are consistent with the real requirements of tasks.

Chapter 13 reviews the main arguments in the book to show contribution that HTA can make to human factors and human resource interventions.

Chapter 14 contains notes relating to various issues that have arisen throughout the book.

While each of Chapters 7 to 12 introduces human factors and human resource management concepts, these topics are not dealt with exhaustively. They are explained only to the extent that it is necessary to demonstrate the contribution that HTA can make. The reader must look elsewhere for a comprehensive review of these issues.

Chapter 1

Task analysis, concepts and terminology

Tasks are concerned with what people do when they work to achieve objectives. Although the word 'task' is used frequently by people in normal discourse, it is not a rigorously defined scientific term and it is certainly not used consistently with respect to task analysis. The aim of this chapter is to explore the concept of 'task' with respect to HTA.

It will review issues of systems thinking and systems ideas that underlie skills. Tasks are considered in system terms. Tasks relate to the system's goals, the resources made available to people and the constraints that must be observed in meeting goals. Task analysis is seen as a process of investigating system factors which influence human performance.

Introduction

HTA was developed by two industrial psychologists, John Annett and Keith Duncan, in the 1960s[1]. Their intention was to prescribe a method of examining work which combined describing *human activity* with understanding the *purpose* of work in terms of the organisations and systems in which it was undertaken. Their method of analysis was also intended to provide a practical way of identifying problems which could then be addressed by human factors solutions. These ideas stemmed from two principal considerations - how systems theory was used to understand complex systems and how the ideas of systems theory were used to understand human performance. In developing HTA, they combined a number of basic ideas from existing task analysis methods.

Systems thinking[2]

A system is a *complex grouping of interrelated parts*, and can include human beings and machines. These parts interact to serve a *purpose*. For example, 'transportation' can be treated as a system to move people or objects from one place to another. It has a purpose and it has components that interact in a suitable way to realise that purpose.

Systems, subsystems and their interaction

Systems may be broken down into *subsystems*. A transportation system, for example, would include, among other things, subsystems which provide vehicles for transporting and subsystems for obtaining the things to transport. These subsystems also enjoy the properties of systems in their own right, so they too may be broken down in terms of subsystems. Thus, the function of providing vehicles for transportation would have a function of procuring new vehicles and a function of maintaining their availability through maintenance (Figure 1.1).

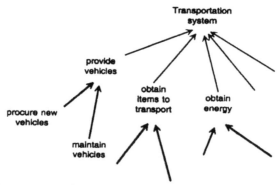

Figure 1.1   Systems and subsystems in a hypothetical transportation system.

[1] This and other notes are dealt with in Chapter 14.

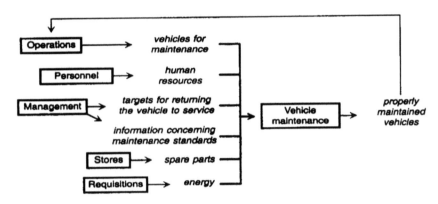

Figure 1.2   Inputs and outputs to related subsystems.

To achieve the purpose of the overall system, subsystems need to *interact*. This means that together they produce something that they could not produce separately. This interaction occurs through subsystems having *inputs* and *outputs*. The inputs to a transportation subsystem of maintaining vehicles includes vehicles needing maintenance, spare parts, human resources, information concerning maintenance standards, targets for returning the vehicle to service and energy, such as fuel and electricity, in order to use equipment and test results. With these inputs the maintenance can take place according to requirements and provide the output of a properly maintained vehicle. These inputs arrive at maintenance as outputs from other subsystems, such as operations, personnel, management, stores and requisitions (Figure 1.2). The output, namely properly maintained vehicles, becomes the input to other subsystems such as the operational subsystem for moving cargo from one location to another.

Just as examining a system will reveal subsystems, the main system being investigated is itself part of a wider system. This means that it will be subject to the values of the wider system. It will also receive inputs from and provide outputs to other systems. Thus, if the items being transported were food, inputs would include the food to be transported, its destination and a specification of the conditions for storage, including the permissible times for transportation. Each of these inputs would need to be satisfactory and the transportation system would need to take proper account of the inputs. If food failed to arrive in a satisfactory condition this could be caused by a failure of the storage conditions, failure of the initial production of the food, or inappropriate specifications being given for storage. Inappropriate specification could have been given due to a confusion regarding the food being transported or an incorrect technical judgement being made by food scientists.

System control

An integral part of systems thinking is the *control* and *regulation* of behaviour. This is often done through the mechanism of *feedback*. Feedback is where information from something we are trying to regulate is used as an input to help determine what should be done to make a suitable adjustment. For example, maintaining the speed of a motor car is achieved by the driver monitoring the speedometer then causing the car to change speed - by using the throttle, the brake, allowing compressions in the cylinders to slow the car or by allowing gravity or friction to change its speed. Feedback and control is an integral part of all systems, including human behaviour and mechanical and electronic systems. The cue for ordering new stock may come from monitoring stock levels. The requirement for people to retrain or improve may come from feedback about how well they have been performing or from identifying new challenges over the horizon.

A closed-loop system is where performance is affected by feedback. An 'open' system is where adjustments are made to a system without reference to feedback. For example, a criminal justice system where punishment is dictated by the current amount of crime in society is an example of 'closed-loop' control. If the punishment was dictated by some general feeling of retribution and not as a consequence of crime rates, this would be 'open-loop' - that is we might increase the severity of punishment for tax evasion if it was felt that such behaviour was morally reprehensible, irrespective of whether it affected the rate of misdemeanour.

A further important feature is a specification of system components (i.e. subsystems) and how they fit together. For example, air traffic control is a system which includes components for monitoring, collecting, storing and transmitting information, so that there can be a safe, efficient and expeditious flow of aircraft. The flow of information through the system is critical to overall functionality - aircraft make requests in relation to a filed flight plan, this is checked against the current and future demands on the airspace under a controller's remit and the request may be allowed or denied. Thus, if a message is lost or garbled or aircraft infringe on each other's airspace this has to be due to the failure of at least one component to meet system design specifications. A failure to display an aircraft on a radar screen indicates a fault in the radar and not in the radio communication function. The primary aim is that each component (human or machine) is able to fulfil the functions required by the overall design of the system. Thus, a system can only be properly understood in terms of its subsystems by indicating how information, energy resources and physical entities flow between them.

HTA and systems thinking

The links between systems thinking and HTA are apparent. Systems have purposes and tasks have goals. Systems can be examined by exploring subsystems and this can only be done sensibly if effort is made to establish how these subsystems relate to one another, both in terms of location within a hierarchy of subsystems and in terms of information and control between subsystems. HTA examines goals by the process of redescription using subgoals and plans. Systems reside within wider systems and are influenced by these wider influences. There is a danger in any task analysis that attention is focused only on those parts of the system which appear unsatisfactory. Thus, a manager might blame the sales-force for poor sales figures, where the real problem might be that the materials for manufacture have been downgraded, increasing the failure rate of products and so degrading the company's reputation. Task analysts should be aware of the systemic influences that affect a part of the work being studied. Equally, goals reside within a context which can influence performance. Repairing a piece of electronic equipment is not the same task if it is carried out in a warm workshop rather than outside in a howling gale on a North Sea gas platform. Systems thinking has greatly influenced the development of HTA and, indeed, any task analysis method which purports to serve practical ends, needs to be carried out beneath a general umbrella of systems thinking.

Systems thinking in describing human skill

Various approaches to the psychology of skill and training research have exploited these systems notions. These were used by Annett and Duncan to justify the use of the hierarchical principle in task analysis.

Goal directed behaviour

Where designers create tools for people to use, they assume that people will adopt the goals that are set for the overall system in which they are operating, for example, motor car designers assume that people will drive their cars in a reasonably conventional manner to move from place to place; designers of video-tape recording systems assume that people will use their machines to record or play video-tapes. Managers, engineers and designers concerned with developing industrial, commercial or health systems assume that people employed will adopt the goals of the wider system. This is a risky assumption, because people might be perverse or might simply not appreciate what they are supposed to do. So part of the job of human factors and human resource practitioners is to help ensure that operator goals are consistent with system goals - through personnel selection, training and effective management. Task analysis methods must address the goals that people seek or try to attain in the systems where they are employed.

Feedback and control in performance

When a person pursues a goal it entails adapting behaviour to influence a controlled environment in order to attain that goal. A central concept in applied psychology of skill is that of *feedback* as a means of enabling purposeful behaviour. There are a number of different forms of feedback that can affect human activity, including information gained through monitoring limb movement, the feel of how controls respond when they are moved, the operation of tools under control, the consequences of the action of tools on the environment and comments from other people concerning the manner and the outcome of actions. Feedback is necessary when using the simplest of tools. To hammer a nail requires controlling the movement of the hammer through the air such that it will strike the nail cleanly and with sufficient force. The user must see what he or she is doing to control the blow, must feel the weight of the hammer and must monitor how the hammer strikes the nail to determine whether it struck true and whether further striking is required. If these aspects of sensation and feedback are missing, then performance will be disrupted.

Feedback is also fundamental to skill in non-manual tasks. Modern industrial and commercial tasks entail the human operator in a complex system having to work via a computer interface or a control panel. It is the responsibility of the person designing this interface to ensure that adequate feedback is provided so that the operator can regulate his or her behaviour. In hammering a nail, the feedback is natural and felt as a physical sensation through the arm. In the control of an automated system it is not. Unless the designer makes a deliberate effort, suitable feedback may be omitted. To provide effective feedback the interface designer must understand what the operator has to do.

Hierarchical organisation of behaviour

Several models of performance use the idea of hierarchical organisation of skill. The operator is assumed to achieve goals by deploying sub-skills. To make a cabinet, a craftsman uses a number of techniques including, selecting materials, sawing, planing, hammering, drilling, screwing and gluing. Each of these is achieved by organising actions to achieve goals. The craftsman judges what should be done next and selects the appropriate subgoal to be met. If the goal is to plane a piece of wood, a decision is made concerning the extent of the planing. Planing entails coordinating a set of sub-skills including manipulating the plane and monitoring results. Manipulating the plane entails coordinating pressure with movement in such a way as to achieve a smooth result and avoiding tearing large chunks from the wood through being impatient. Thus the goal is sought and accomplished by the operator organising various sub-skills to accomplish this end. The effects of these skills is monitored so that the operator can use this feedback to regulate action or decide when the next thing must be done. Thus, behaviour is goal directed and goals can operate at a number of hierarchical levels via feedback loops.

The principles of goal directed behaviour, feedback and hierarchical organisation have provided system models to account for skilled human performance and these are often used to justify methods of task analysis. Indeed, HTA was originally justified in these terms. However, there is a problem with this approach because different people organise their skills in different ways to accomplish the same task. Moreover, an individual will deal with a task in different ways in different circumstances — as people gain experience they start to work more efficiently and if circumstances change to make a successful outcome more critical, then the person will become more or less cautious and may use a different method. Hierarchical models of skill are useful in helping us to understand and represent behaviour, but using them to account for how *all* operators do or should carry out a particular type of work can be quite misleading.

A more versatile approach to this issue is to consider the task in terms of the demands placed on operators, rather than trying to model how all operators would respond. For this we can turn to the idea of examining human performance in the system context.

Human performance in systems[3]

When considering how people carry out real tasks, we need to understand the context in which they are operating. The context will supply information and provide tools that people may use to accomplish goals. We cannot begin to understand how people make decisions unless we appreciate the information they can use and the opportunities for action they have available. The wider context will dictate the frequency with which this information is presented. A person can develop different strategies for responding when things occur routinely rather than when they are unexpected. The context will provide the wider system goals to which the person is expected to contribute. Therefore, it affects the important values of the system - which activities are crucial and which are trivial, which factors need to be taken into account when making decisions and which factors can be ignored. It also indicates how people interact with other people, with machines and tools, and with plant and equipment, in order to meet their working goals. An understanding of factors governing behaviour is important with regard to each of these considerations.

Figure 1.3 represents a person working within the wider system. Here, the person has been referred to as the 'operator'[4]. In this diagram, the 'system in interaction' refers to the plant or machinery or other people or combination of these things, that the person is trying to influence. Thus the system in interaction would contain a hammer, nail and piece of wood, if the operator is hammering; if the person is driving it would include a car, other cars, obstacles and the road itself. Indeed, the system in interaction could contain all sorts of things that are not apparently obvious. The operator interacts with this system in order to carry out the task. The system in interaction is represented to the operator via sight, hearing, touch and smell. This representation can be thought of as the task *interface*. The driver's performance is also subject to other influences, for example glare or noise from the road. These can affect the efficiency with which

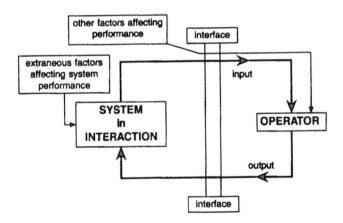

Figure 1.3   The expanded operator-system interaction, showing information and feedback loops.

information is processed, thereby making the driving task more difficult. In principle, the system in interaction with any operator is everything else in the environment. Often it is only as task analysis progresses that we are able to assess which parts of the environment need to be taken into account.

The right-hand side of Figure 1.3 refers most directly to human performance, while the left-hand side is the amalgam of factors that influence the signals that the operator experiences. The interface is the way the world is represented to the operator. Sometimes we are working within a domain where the interface can be altered, such as in the design of computer screens or instruments in vehicles, or even in the way in which we might require people to communicate with one another. In other cases, we cannot affect the design of the interface - the operator must use what is provided by nature.

We can expand the operator side of the system by speculating about the information flows between the operator and the system. Figure 1.4 shows how information from the world outside may be considered as an *input* supplying the operator with basic information upon which to make *decisions*. Decisions are concerned with planning responses to events and controlling these responses such that the system's goals can be met. When an *action* is selected it must be *controlled* to ensure that it does what is required. *Action feedback* indicates whether the control action needs to be regulated in any way. For example, to switch on a light we need to control the movement of the hand towards the switch and then indicate when pressure needs to be applied. *Control feedback*, tells the operator how the control device has been affected, for example whether the switch or lever has moved as intended or whether a computer key was properly pressed. *System feedback* tells the operator how the system has now changed as a

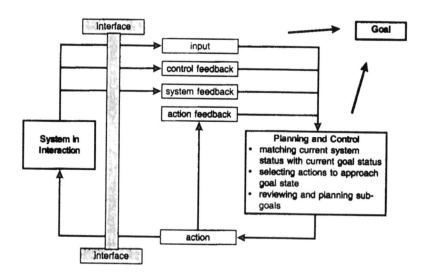

Figure 1.4   The human-task system expanded to show additional elements of human performance.

consequence of the control action. In tasks, the operator determines what needs to be done by evaluating information about the system from the *input* and *planning* a suitable response. A successful control action is one where an appropriate action has been selected, the action is executed successfully, the controller works as intended and the system behaves in the manner expected. This basic model can be seen to apply to nearly all situations, including social interaction as well as more physical activities.

The examples given here refer to physical actions. The same principles can be applied in understanding interaction between people. Giving a visitor the directions to a house entails monitoring that an appropriate direction is given, monitoring that the visitor has appeared to understand the directions given and then confirming that the visitor actually arrived.

The significance of system feedback, in particular, is critical to skilled performance. A task analyst needs to understand something of the dynamics of the system in interaction in order to appreciate how system feedback can affect performance. In controlling a large continuous process plant, such as a petrochemical or power generation plant the operator must often wait for some hours before being confident that an action has had the desired effect. For example, to raise the temperature in a vessel by 10 degrees may require that one of the feed materials is first heated to a higher temperature. A decision must be taken to select which action is appropriate and by how much the controlled parameter should be changed. When the change to control has been affected, the system has to be monitored to ensure that this action remains appropriate. Because of the inertia of such systems - they are big with lots of pipes and vessels to heat up - it may be some hours before the consequences of the action can be judged. Then, if it is then seen that the temperature has risen too much, it may be too late to correct the action. This is why effective planning of actions, including decision-making tasks, is so important.

This phenomenon of *slow response systems*[5] is well known in engineering situations. It also happens in many other contexts. In organisations, problems often occur because communications are delayed. An important message for staff to change practices may have been issued too late for adverse consequences to be avoided. Doctors and nurses in intensive care often cannot afford to wait until a prescribed treatment has run its course; they must continually monitor the patient to detect any signs of risk and predict future trends in order to judge the consequences for the patient. Understanding something of the system in interaction helps the task analyst understand the tasks and skills that must be brought to bear in dealing with system events. To what extent must the operator, in a particular system, simply respond to events in a stereotyped way, how much monitoring should be done, to what extent can actions be selected with confidence according to an effective diagnosis and to what extent can the operator only make informed guesses, then plan for a variety of contingencies? Task analysis cannot be undertaken solely by reference to what the operator does. Considering the wider system often provides information about the operator's motivation and helps to indicate the significance of errors.

Justification of HTA in systems terms

A task is something that people do, but to define the word 'task' more precisely, it is helpful to consider the system in which the task resides.

The task system

This 'task system' represents an interaction between a human operation and the operator's environment in order to achieve a required goal. Figure 1.3 is modified in Figure 1.5 to show how the operator uses information in decision-making to select and carry out control. The set of things in the grey box amount to the 'operation' - this is the closest that HTA gets to making statements about how behaviour is organised. How behaviour is actually organised is a question for cognitive psychology, but we may presume that such behaviour includes the functional stages of *perception*, in which sensations detected in the environment are translated into a form which the operator can then use to *make decisions* in order to exert *effort* to manipulate controls. By using the idea of an *operation* as a unit of behaviour, the analyst can avoid the intricacies of attempting to model human behaviour in a way that would satisfactorily account for all of the different ways that people could achieve the same goal in similar circumstances. The concept of 'operation' has always been central to HTA. Annett and Duncan described it as ....*any unit of behaviour, no matter how long or short its duration and no matter how simple or complex its structure, which can be defined in terms of its objective*[6].

The operator is assumed to have a goal which directs his or her behaviour. For the overall system to work, the operator's goal needs to converge with the system's goal (GOAL). The diagram has also added the influence of the *environment*, both in affecting the system in interaction and affecting the human operation. Environmental factors can substantially alter how the system works. For example, if a car is driven on an icy road, it will affect the handling of the car. The actions that the driver would carry out in safer circumstances, may be inappropriate here. Thus, for goals to be met to a satisfactory

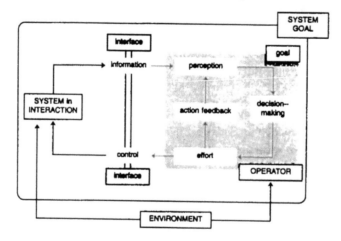

Figure 1.5   The human-task system showing the influences of the environment.

criterion the human operation - the driver's behaviour - must be sufficiently skilled and adaptable to cope with the problems created by the road conditions. Equally, if the environment caused an industrial plant to operate differently, then the operator would need to adapt in order to meet goals. A sales person in a book shop may need to become better acquainted with Cajun cooking in order to advise customers, should a television programme create a Cajun cookery boom. The environment can also influence the manner in which the operating behaviour is carried out. For example, if the environment causes a person to become tired or stressed though excessive demand, or causes illumination to make vision difficult, or mean that the consequences of making mistakes have become costly, then human performance will be affected.

For most purposes in HTA the analyst can limit references to human behaviour to the operation that the person carries out to meet a goal. In some circumstances the analyst may feel there is benefit in speculating further about the behaviour involved. This is fine because such speculations can be carried out in the context of the emerging HTA, as we shall discuss in Chapters 2 and 4.

A possible source of confusion in HTA is that the words 'operation' and 'goal' are sometimes used interchangeably. This is because for every goal that a human being has to carry out, there is a unit of behaviour which enables that goal to be met. Thus, 'operate toaster' is first used to specify the goal to be achieved. Then it is used to represent the behaviour necessary to operate the toaster. If the analyst judges that this operation is likely to be carried out successfully, then no further analysis is required. If, however, the behaviour is not judged to be satisfactory then further analysis is required - the goal is broken down into its subgoals and their plan. The same consideration is then given to each of the subgoals, that is, judging whether their corresponding operations will be carried out satisfactorily. Writers on HTA have used these terms interchangeably with little consequence. However, the distinction is worth making because it underlines the strategy of the task analyst which will be dealt with more fully in Chapter 2.

Task

If we were to remove the greyed area from Figure 1.5, we would be left with that part of the system with which the operator must cope in order to meet the system's goals. This set of demands includes the system's goal. It also includes the information that will be available to make decisions, including making judgements and using feedback. It also includes the control facilities available to change circumstances, including control and communications devices and other people to whom messages can be given. It also includes the environment which affects both the signals to the operator and the conditions under which the operator must perform. These elements, then, represent the *task* that the operator must try to carry out. Thus, a task can be treated as a set of things which comprise a *system goal*, *resources* for accomplishing the system goal, including *information* and *controls* and a set of *constraints* on how the goal may be achieved using these resources.[7]

This view treats the concept of 'task' as a challenge facing the operator - a problem that must be solved in order to attain the system goal. This makes sense when we consider a practical example. Preparing multiple copies of a document can be achieved in several ways: photocopying if a photocopier is available; multiple printing if the document file exists on a computer; photographing and printing if only a camera is available; writing out the extra copies by hand if only a pen and paper are available. Clearly resources affect the freedom of action. A sense of urgency - for example, the document must be produced to a high quality by lunchtime - might warrant taking the master document to a copy-shop, provided there is the means to pay for this. There are often many ways in which operations can be carried out, but the choice of which method to follow is limited when the available resources are known and the choice is further refined when other constraints are known. The operator's task is to work out how to attain this goal given these resources and constraints - indeed, how to plan the use of available resources to attain the goal.

A further aspect of this interpretation of 'task' is that, as far as the operator is concerned, the task relates to his or her immediate experience. From this perspective a person experiences the world through information and control artefacts. Information can provide knowledge about the task including current status information which will indicate how the system in interaction is working, how it has been affected by controls and how it has been affected by the environment. It can also provide 'noise' in the system which could cause distraction or otherwise interfere with performance. The operator will experience an amalgam of these things and not have the luxury of knowing what variations have been caused by what. A designer has the opportunity to make some features prominent and subdue others so that the operator is better able to decide what is relevant and what is not. Part of the task analyst's job is to identify which features of the task are most useful to the operator in order to show which should be most prominent or to demonstrate where lack of suitable information can lead to error.

Task analysis

Having now suggested what a task is, we can consider what 'task analysis' is. The account, so far, has indicated how performance of an *operation* is affected by a wide range of factors, including the events generated by the system in interaction, the way in which these events are represented to the person trying to control the system and the environmental influences on all parts of the system. In human factors work, or management generally, we are concerned with actual or potential mismatches between the system's goal and the operator's goal. What has caused inadequate performance? What can be done to eradicate inadequate performance? To answer these questions the task system must be explored. This process of exploration is *task analysis*.

Hypotheses in design

One of the challenges of task analysis is that the analyst is attempting to make assertions about human performance, often in novel situations, without possessing categorical evidence of how behaviour operates in a given set of circumstances. Psychologists often investigate behaviour by carrying out controlled experiments on simplified tasks which allow them to limit the range of influences affecting the situation. This is done to enable them to draw inferences about why people behave as they do. In the situations where we carry out task analysis we must accept the task complexity that is presented and we have little control over environmental influences. It means that we can never be absolutely certain about conclusions reached while the task analysis is being carried out. This is in no way a deficiency of task analysis, task analysts or human factors. It is the same for anyone engaged in trying to predict how lots of complicated factors interact in ways that have not been experienced.

This uncertainty means that any outcome of a task analysis process, be it identifying a potential problem or generating a method to improve performance, must be treated as a *hypothesis*. Hypotheses, such as those concerned with improving training or reorganising a work team in order to secure performance benefits need to be evaluated. This may be done formally by a trial, or informally by making a change and then monitoring the outcome. Figure 1.6 shows the cycle of task analysis, hypothesis generation and evaluation that is implicit in all task analysis interventions. Indeed, this cycle is implicit in all managerial and design interventions. People in authority in systems may deny the truth of this and assert that it is possible to make assertive statements about the certainty of different hypotheses. Indeed, consultants may also adopt a stance of certainty to impress a client. But these are merely adopted stances and are wrong in principle.

There is no need to be defensive about this. It in no way diminishes task analysis methods. Indeed it emphasises their importance, because by adopting methods to examine tasks which are systematic and rigorous, hypotheses are more likely to be successful without excessive rework and adjustment. Moreover the cause of subsequent problems will be more easily pinpointed. Indeed, task analysis methods should be part of the armoury of managers to enable them to deal more effectively in managing uncertainty.

A further insurance against hypotheses failing to be effective is that human beings are intelligent and adaptive and that this feature should be exploited in all system design. If we are designing a rigid structure, such as a bridge, then suggesting inappropriate materials or substructures that are insufficiently robust to withstand forces, will lead to

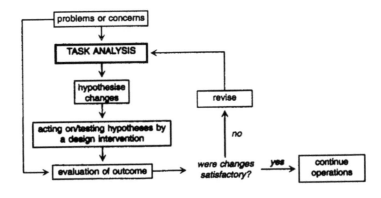

Figure 1.6   The hypothesis and test cycle

errors that must be rectified. In designing for human beings, we can rely on their capability to deal with the unexpected, but we need to take steps to allow them to be flexible and effective in this way.

Design options and constraints

A further attitude to task analysis that should be dispelled is the view that there is only one correct answer to a problem investigated by task analysis. It is usually possible to bring about the same improvements to an operating situation by improving training, interface design, job-aids, reallocation of duties, modification to the wider system in order to influence the scheduling of events. Indeed, combinations of these may also serve to achieve the same ends. The analyst must choose an appropriate combination of compatible interventions. This is not as complex as it sounds. It simply requires a breadth of perspective.

Any design activity is concerned with identifying options and then using constraints to help reduce the options that can realistically be considered. In designing an aircraft, for example, the designer may consider a range of engines capable of attaining the required thrust. However, this list can be reduced when other criteria such as cost and weight are also taken into account. Task analysis requires the same perspective. If the analyst has ascertained that the company will not recruit new staff and will frown upon excessive interruptions to production incurred by changing the physical working environment, then solutions such as team reorganisation, training, job-aids and job-redesign may be sought. However, if the analyst cannot see a satisfactory solution being crafted from the apparent options, then he or she must notify the client that improvement is impossible whilst current constraints apply.

There are *constraints* in all systems which affect how resources can be manipulated in order to achieve goals. Constraints include physical constraints - some things are too hard to do, some things give rise to other consequences which should be avoided. Some constraints are imposed by management to ensure goals are achieved without using too many consumables that could affect profitability- such as too much paper or too much power. One such set of constraints concerns compliance with the law or sets of standards. In the manufacture of pharmaceuticals, for example, acceptable manufacturing practices are enshrined in the manufacture of ethical drugs, which means that constraints are imposed on a company with regard to the methods they might adopt in manufacturing. Equally, health and safety legislation might result in certain safety procedures being essential. Some constraints stem from the operator's humanity or sense of community. For example, an operator may limit activity that could cause problems for a new shift.

This need to appreciate design options and constraints is important if task analysis is to serve the needs of a client in a practical way. Thus, the analyst cannot be content with simply knowing about human performance; he or she must become acquainted with the systems being operated. A broad understanding of how systems are managed is important in learning to be an effective task analyst, as is acquainting oneself with the organisation being investigated.

Some terminology so far

As we start to discuss task analysis and HTA in a technical way, it makes sense to adopt some common terminology. The following list begins by describing how to refer to people engaged in the task analysis process. It then considers terms used or alluded to so far in the Introduction and Chapter 1.

People engaged in the task analysis process

Operator. The word *operator* will generally be used to refer to the person who is carrying out the task. Where it seems more appropriate, I shall sometimes use the word *user*.

Analyst or Task analyst. The approaches discussed in this book are relevant to a wide range of professionals including managers in all disciplines, engineers, management support officers, occupational psychologists, ergonomists, human factors engineers, personnel officers, information technologists, training officers, safety analysts, inspectors, work-study engineers and technical writers. For clarity and consistency, I shall often use the general term *task analyst* to refer to the people engaged in the processes of analysing tasks.

Client. An analyst is likely to be somebody working on behalf of someone else, possibly a manager or engineer. The word *client* is used to identify the person for whom the task analysis is being done and who has responsibility for the task and system being analysed.

Informant. The *informant* provides information to the analyst. The informant will be an agent of the client and may be briefed to provide whatever information the analyst requires, including system's knowledge and operating practices. The informant may also include demonstrate the task to the analyst.

Summary of general task analysis concepts and terminology as they relate to HTA

Task system. The *task system* is the system in which the activity of interest resides. It includes the system in interaction, such as a plant or a working team. It also includes the environment.

System goal. The *system goal* is a statement of what the system is required to achieve. It follows that, since the operator is part of the system, he or she must carry out actions which are consistent with the system's goals. Goals are expressed as an *instruction* or imperative, given in the form *verb-action* or a combination of such instructions.

Criteria. The word 'goal' is used to signify a target, but does not necessarily contain all of the information necessary to specify when the target has been reached. As analysis progresses, it is important to accumulate the criteria with which performance must comply. Often, the detailed criteria only emerge as analysis progresses.

Operator's goal. People direct their behaviour towards what they believe needs to be achieved. This is the *operator's goal*. Much of task analysis and human factors work is concerned with ensuring compliance between the system's goal and the operator's goal.

Task information. To achieve a goal the operator must use information to provide cues to prompt action, information to make decisions, and information to provide feedback to regulate action.

Operating resources. In choosing actions, the operator must use the *resources* available. These resources include anything that may be manipulated to affect the system to achieve goals. They may include: controls which have been installed for the operator to use; *raw materials* that can be directly manipulated, possibly by the use of *tools*; *other people* who can be influenced by the operator in ways that will affect system performance.

Interface. The representation of information and resources to the operator is called the *interface*. It is the interface between the operator and the wider system that the operator is trying to manipulate to achieve the goals.

Operating and design constraints. As task analysis is intended to serve a practical end, any *design options* made to improve matters must take account of *design constraints*. All organisations operate within financial and legal constraints. There are also technical and ethical constraints on design options concerned.

Task. A *task* is regarded as a problem to be solved or a challenge to be met. A task can be regarded as a set of things including, a *system's goal* to be met, a set of *resources* to be used, and a set of *constraints* to be observed in using resources.

Operation. The *operation* is that which the operator does to achieve a goal. It is the process of using resources to affect the system being controlled. An operation is a reference to the behaviour that the operator undertakes to achieve the goal.

Hypotheses. Where a design suggestion is made to overcome performance problems, it cannot be assumed that it will work until it is demonstrated. Therefore, insights and design suggestions, though an essential part of task analysis, must be treated as *hypotheses* until evidence is gathered to demonstrate their effectiveness.

Task analysis. Task analysis is treated here as the process of obtaining information about a task in order to generate hypotheses concerning sources of inadequate performance or about designs that will make things better. Task analysis entails examining performance of operators and examining the characteristics of systems that can influence performance. It also entails exploring practical constraints on design options to ensure that suggestions are realistic.

Summary of HTA concepts

The concepts of HTA will be discussed and illustrated more fully in the chapters to come. For the present, the following ideas have been introduced so far.

Subgoals. A *subgoal* is part of a wider goal. A central strategy in HTA is to redescribe goals in terms of their subgoals.

Plans. A *plan* in HTA describes the statement of the conditions under which each of a set of subgoals is undertaken to achieve their common *superordinate* goals.

Redescription. *Redescribing* a goal is the process of working out how a goal can be represented in terms of a set of subgoals and their plan, such that the set of subgoals and their plan is equivalent to the goal being redescribed. A *redescription* of a goal is the set of subgoals and their plan which accounts for the activities that need to be carried out to achieve the goal.

Criticality. The achievement of goals or subgoals could be critical or trivial. If a goal is trivial it does not matter whether its operation is carried out particularly successfully. Mistakes may be an irritant, but still may be tolerated. If a goal is critical then it must be carried out successfully, according to a performance standard which can include production and error rates.

Stopping redescription. Where errors are trivial, there is little need to put them right, therefore, the costs and effort of redescription can be avoided. By using a stopping rule in HTA which includes consideration of the *criticality* of performance and the *likelihood* of the operation being carried out properly the analyst can focus on those areas of the task that are of most concern.

Concluding comments

This chapter has focused on the broader issues of task analysis as well as setting out some of the ideas fundamental to HTA. The importance of systems ideas is stressed. Systems thinking is an important notion in HTA and other task analysis methods. It emphasises that human performance in practical settings must serve the needs of a wider system, therefore human performance must be linked to system's goals.

HTA, as with many task analysis methods, derived from systems ideas as they were applied to explaining human skilled performance. Caution is advised here, because it is apparent that different people adopt different methods for carrying out the same task. Moreover, individuals commonly vary their strategies for dealing with similar situations as circumstances change. This means that it is rarely possible to offer a single account of skill to represent how people, in general, carry out a particular task.

Introduction

In this chapter, the strategies for examining tasks, using HTA, are developed. The chapter
will first consider three main ways in which goals or operations are examined. It will
then show how these are used within a framework for task analysis. By working within
this framework, a task is explored in progressively greater detail until it is understood
sufficiently. Hierarchical diagrams are an effective way of representing the *product* of
this examination. However, the real power of HTA resides with this process of
investigation.

The main strategies of HTA

Strategy in task analysis is essential to making progress. Figure 2.1 illustrates the main
cycle of decision-making used in HTA. The overall goal is stated. If there is no concern
for the manner in which it is carried out, then further action is unnecessary. If there is
concern, then the analyst can first examine the *human-task interaction* and the *operations*
underpinning performance, in order to establish whether these problems can be identified
or whether design solutions can be proffered. If no hypotheses are forthcoming, usually
because the goal is still at too coarse a level of description for sufficient insight, then
redescription, into subgoals and their plan, will be attempted. How this can be done

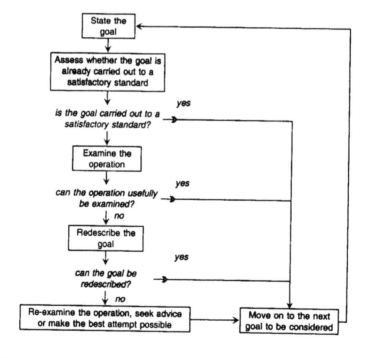

*Figure 2.1   The basic cycle of decisions during task analysis. This is expanded in Figure
2.3.*

will be described later. However, there are occasions when satisfactory redescription is not possible. In these cases, the analyst is forced to reconsider or seek advice. Then the analyst follows the next goal for treatment in the same way. By following this process a task hierarchy of goals and plans emerges. The characteristic HTA diagram is simply the product of this cyclical process.

Three main analytical methods are employed in Figure 2.1. The first is concerned with deciding whether the goal warrants examining (1). If the goal warrants examination, then this is done by one of the other two methods. First there is examination of the operation by considering how the operator and the task interact (2). This entails the analyst making judgements about weaknesses in human performance or the task conditions that might cause difficulty. These considerations can also lead the analyst to suggest specific improvements. The second method of examination is *redescription* (3). Redescription is the process of reconsidering a goal in terms of its subordinate goals and their plan. In many respects this is the defining method within HTA and it is the aspect that will receive greatest attention throughout this book. However, it is stressed that HTA is best seen as a *framework* for task analysis that moves the analyst systematically between these three main considerations. This framework will be expanded upon shortly.

(1)   Deciding whether to examine a goal

Since task analysis is a practical activity, there are often time constraints on how it is carried out. For example, it is often essential that a task analysis is completed in good time so that decisions can be acted upon. Also, there may be a limit on how much access the analyst can have to personnel and the workplace. For these reasons effort in analysis should be directed where it is most essential. Thus, before examining a goal, the analyst should judge whether the effort is necessary.

Examining goals is pointless if the goal is not important. When analysis commences, the main goal to be analysed invariably warrants examination, otherwise the project would not have been initiated. However, when subgoals are examined the analyst needs to focus on the most critical areas. Judging whether a subgoal is worthy of further investigation entails a form of *cost-benefit analysis*. It is concerned with the *risk* entailed in the operator carrying out the operation. This involves the likelihood of the operator committing an error and the consequences that would arise when an error occurs. In HTA, this was expressed as the $P \times C$ rule, where P stands for the *probability* of inadequate performance and C stands for the *criticality* or *cost* of inadequate performance.[8]

Generally, P and C are estimates made by the analyst in conjunction with the client or the client's agent, and often the judgement is made intuitively by the analyst in order to make progress. It is rare for P and C to be precisely quantified, though in some circumstances they could be. In a repetitive assembly or inspection task, for example, data could be collected that would show the frequency of error and the costs of rework or the costs of disposal of unsatisfactory items or the costs of replacing faulty goods that had been sold. Generally, though, the analyst relies on estimates. Their product $P \times C$ is simply a convenient shorthand for combining these two factors. Thus, if the estimate of C tends to zero, the product P x C will tend to zero. In some workplaces, for example, staff are required to keep logs as a method of prompting them to keep an eye open for unusual events. It may be the case that log entries are never checked and poor log entries go unnoticed. Therefore, C would be low, approaching zero, so P could take any value and still the product P x C would tend towards zero. In other situations log-keeping is important, because the data contained in written logs are crucial to evaluating production problems, so this conclusion would not apply. Accidents in public transport systems or nuclear installations may be comparatively rare, but when they do occur their consequences can be catastrophic. Therefore, despite the extremely low value of P brought about through well-engineered systems and good training, the unacceptably high value of C means that there is no case for complacency.

(2)   Examining Operations

Examining the human-task interaction requires the analyst to focus upon a particular goal (or subgoal). The analyst has considerable freedom to choose how this is done. Indeed, this is where analysts are invited to draw on other task analysis methods and use their own experience. There are many different methods and perspectives that can be used here, although, in many situations the analyst will be content to make an intuitive or expert judgement. Four common human factors strategies may be employed:

1.   to model the psychological processes underpinning the operation;

2.   to take advantage of the common similarities that exist between actual operations even from widely differing contexts;

3.   to treat each operation of concern to a systematic appraisal using, for example, a checklist;

4.   to subject the operation to a process of further data collection using a specialist method.

Modelling behaviour

Modelling behaviour entails trying to understand *how* people accomplish goals by making reference to models of human performance. This approach featured centrally in Annett and Duncan's original ideas for HTA where they employed a simple 'information-processing' model of skilled performance to help the analyst gain a speedy insight into the possible problems that an operator might encounter. They suggested that an operation (i.e. what the operator does), should be thought of in terms of 'input', 'action' and 'feedback'. That is, competence at an operation implies an ability to collect information (*input*) pertinent to the execution of the task; an ability to carry out the *action* selected in order to move towards the goal state; and the ability to monitor appropriate *feedback* to determine whether the action is being executed correctly and is appropriate for dealing with the goal in question. This is illustrated in Figure 2.2 (a)[9]. There is no explicit *decision-making* component in this scheme, but use of feedback to regulate action to meet the goal implies the necessary planning and decision-making skills. This sort of scheme enables the analyst to consider, systematically, the likely sources of human error in the conduct of an operation. If an *input* or a *feedback* weakness

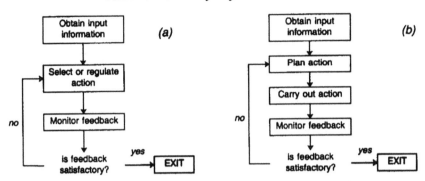

Figure 2.2   A simple information-model of an operation. (a) shows how the operation is represented using input, action and feedback. (b) replace 'action' with planning for an action and executing the action.

is suspected, the analyst would be directed towards considering the display of information to the operator, or training in the discrimination, organisation and interpretation of signals. If an *action* weakness is suspected, the analyst could be directed to consider equipment redesign or training.

A modification to this scheme is to distinguish explicitly between the *planning and decision* making components and the execution of the *action* itself - see Figure 2.2(b). This could help the analyst focus on very different aspects of performance. Problems with planning and decision making are very much concerned with operator strategy and knowledge, whereas problems with action could be concerned with motor skills, physical fitness and inappropriately designed controls. Indeed, identifying a problem as one that is concerned with planning and decision making can open up other alternatives for representing cognition.

These simple models are *information processing models* of skill. There are many such models in the literature, which vary in detail. Some explicitly mention short- and long-term memory to prompt the analyst to think about performance problems concerned with forgetting crucial information either about what has just been happening in a task or remembering important knowledge about the task that the operator has forgotten to apply in this instance. These schemes also vary in the extent to which their authors present them as scientific models of behaviour or simply useful tools to help an analyst examine a problem in a systematic way.[10]

Identifying similarities between operations and goals

Another method the analyst can adopt is to try to identify characteristics of a current operation and relate this operation to others encountered - insights or solutions that were seen to be appropriate elsewhere might be promising on this occasion.

Undoubtedly, similarities do exist between operations from different domains. Recognising and exploiting these is one of the key elements of experience of task analysts. Chapter 3 discusses commonly occurring plans and Chapter 6 discusses a number of tasks in different domains. Both of these Chapters show how similar task elements arise in different contexts. For example, many tasks rely on an operator *monitoring* a system to detect if and when its conditions go out of specification. Examples include people operating automated industrial plant, supervising transportation systems and nursing in intensive care. Each of these domains is very different. However, in all cases, monitoring requires the operator to know the parameters to monitor, knowing target values and knowing the tolerances outside of which an observed parameter must not be allowed to go. These operations also require the operator to be conscientious and monitor routinely and reliably. Knowing these facts about monitoring can alert the analyst to a number of potential practical issues. In this way the analyst might quickly pinpoint a source of difficulty.

Similarities between operations may be exploited in formal classification schemes or they might simply be the result of an analyst's experience. For example, tasks concerned with dealing with complex systems, where operators must monitor and maintain system status, all entail combinations of the following standard operations:

- Monitor for problem
- Detect potential problem
- Diagnose problem
- Make system safe
- Compensate for problem
- Rectify problem
- Recover from problem

These similarities can help the analyst see how to redescribe such tasks or they can help pinpoint concerns.

Checklist approaches

By 'checklist approach' is meant the strategy where an analyst subjects operations of concern to systematic scrutiny using a list to guide these considerations. A good example would be where the analyst was concerned to ensure that the environmental ergonomics of a workplace were satisfactory. For instance, the analyst might wish to establish whether the operator was subject to any extremes of heat, light, sound or draught that might adversely affect performance. Operations of potential concern would be systematically subjected to a battery of environmental measures and these would be recorded against the operation identified through the HTA. This process has not entailed any modelling of underlying process, but it has provided a systematic set of data for later examination[11]. There are many approaches such as this where the analyst might wish to be systematic in recording data as part of a wider comprehensive audit, for example.

Other methods of data collection

In several cases, the analyst might wish to explore the detail of an operation further by using an appropriate method of data collection. This could provide insights where less stringent methods employed within the HTA have failed.

One such example would be the collection of verbal protocols[12]. The analyst might encourage the operator to verbalise his or her strategy in carrying out a difficult task, record this speech, then examine it later on. The verbal protocol could be recorded concurrently as the task is carried out, or it could be recorded afterwards, relying on the operator's memory or allowing the operator to follow a video recording of what took place. In this way, the analyst may gain useful insight into the operator's strategy, motivation and justification for action. This process can help identify useful task knowledge that could be used to train people, or it could provide evidence of the need to modify the information available to operators during the task. One important outcome is that it could show the analyst how further redescription of the task could be accomplished.

Verbal protocol analysis is but one formal method that can be employed in order to collect data about operations. Other common techniques include 'link analysis' where the analyst records the extent to which the operator makes use of different artefacts in the task, including instrumentation, communications devices, other people and records[13]. Identifying common links can point to ways of reorganising the workplace. Identifying common patterns of using links can point to the skills and strategies that people use and, so, provide insights for further redescription of the task.

Another useful method is the 'withheld information' technique[14]. This simple, but powerful method is helpful in understanding how people diagnose situations in the face of uncertainty. The analyst must prepare an information sheet containing the set of information that the operator could use, but then withhold this information until the operator explicitly asks for an information item. By recording the order in which information is requested in reaching a solution to the problem, the analyst gains insight into the operators strategy, the information upon which the operator depends and the types of error the operator makes.

The three methods mentioned here are all reasonably straightforward. There are other methods, several of which are quite practical, but some such methods are impractical because they take too long, require too much preparation or are unlikely to provide much practical insight. Choice of such methods must be justified with regard to the purpose of the task analysis intervention.

(3)   Examining goals by redescription

If exploration of the human-task interaction proves difficult, unsatisfactory in generating useful hypotheses, or the analyst feels that he or she has not yet gained a suitable grasp of the issues involved then the other way of exploring a goal is by redescription[15]. One reason that the analyst might reject jumping too quickly into relying on an examination of the human-task interaction is that he or she is, as yet, insufficiently familiar with the contextual constraints that affect operator performance. There is a real danger in assuming an operation is understood simply because it has familiar characteristics.

A goal is *redescribed* in terms of a set of subordinate goals and an organising component known as the 'plan', which specifies the conditions under which subordinate goals have to be carried out to meet the system goal in question. The numerous examples of HTA in this book demonstrate redescription. Strategies for redescribing plans are discussed in Chapter 3.

HTA - a Framework for Analysing Tasks

To emphasise the manner in which HTA provides a framework for task analysis, Figure 2.3 shows an expansion of the process set out in Figure 2.1. This contains the same general flow of task analytic activity but sets out the process in greater detail. This will now be described.

1.   Identify the main goal to be analysed

A first step in analysing a task is to identify and focus upon the main goal of the analysis. Generally a client will present the problem to the analyst in some way. However this is done, analysts are advised to work out for themselves the best place to start, because the client's concerns can be misleading.

A task analysis project is generally initiated because a manager (the client), has a concern. This concern may be prompted by a feeling that human factors could be improved generally and therefore a general review is warranted. It might be that the system is under development and a systematic approach to supporting human performance is required. It could be a response to an actual incident, such as an accident or a near miss. It could be a need to comply with an outside inspector in order to show that appropriate standards are being observed within the company.

A client might represent the project to the analyst in terms of a general audit to be carried out, in terms of a problem to be solved, or in terms of a solution to be implemented. A general audit gives the analyst a reasonable degree of freedom concerning where to focus. A problem to be solved can be misleading if the cause of the problem resides outside of the immediate area of the problem itself. Where a client represents a project in terms of a solution to be implemented, the analyst needs to be especially careful because the solution may not deal effectively with the problem that initially concerned the client.

Perspective of the analysis

In Chapter 1 a distinction was drawn between the *system's goals* and the *operator's goals*. Different projects warrant different perspectives. If the project is concerned with designing for human factors, then the task analysis should take the system's perspective and explore what people would be required to do to meet the wider system's goals. If the analysis is undertaken in response to a problem, such as a near miss, the analyst

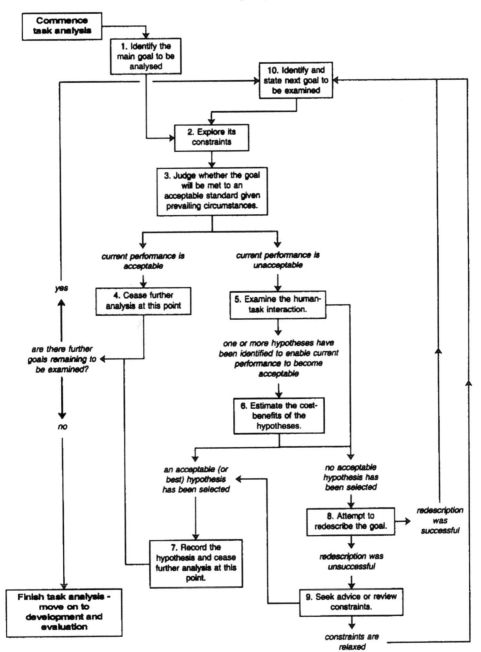

Figure 2.3   The cycle of task analysis decisions. This is an expansion of the diagram in Figure 2.1. Working through this framework enables the analyst to generate the hierarchical task structure characteristic of HTA.

might need to consider the system's goals in order to establish the constraints on how things should be done, and also the operator's goals in order to establish how the task was actually carried out. This would provide a basis for identifying divergence between operator performance and system requirements.

Often there is a need to examine operator behaviour in order to identify strategies that can be taught to others. In these cases, expert operators would be identified; the HTA would be conducted initially from the perspective of the system's goal and would then proceed by considering the goals of the task experts. It is important for the analyst to be aware of this perspective, and it is important to be flexible when working through the analysis.

Focusing on the task

Finding the right level to start analysis also develops with experience although there are rules-of-thumb and analytical techniques which can help this issue. There are analytical techniques which can help this focus, for example, Where the client has real concerns, such as experiencing accidents, near misses or performance problems which affect productivity, the analyst can focus attention towards these. The *critical incident technique* is a method which can be used to refine this focus by systematically considering incidents[16]. Even without a systematically considering incidents, the analyst can focus upon areas that have been reported as being of concern. In both cases the analyst must expresses a suitable working goal to provide focus for the subsequent HTA.

It is often helpful for the analyst first to consider the inputs and outputs to the human-task system being examined. This means understanding what information and materials flow elsewhere in the organisation and beyond, and what information and materials the current system relies upon[17]. The *outputs* will help the analyst appreciate the importance of the goal under investigation to the wider organisation. This can help later on in understanding the consequences of error. It can also account for problems experienced elsewhere. Understanding *inputs* sets out the information and materials upon which the present system relies. Thus, problems experienced in one place may not be capable of solution unless the area where they were caused is identified[18]. It is good analytical practice for the analyst to start task analysis by becoming acquainted with the wider system in this way. Then, if problems arise in the analysis which might be difficult to resolve, alternatives might exist elsewhere in the system which either allow the problem to be minimised or dealt with in a different way.

Sometimes, as an analysis progresses it becomes apparent that it has been started at too detailed a level of description and so a broader goal should have been considered. It is often a good idea to start the analysis more broadly or, subsequently, put aside the work to date, take a step back to explore a wider goal, then fit the earlier work into this broader analysis.

An example of this strategy is seen in the HTA of a machine packaging operation in a pharmaceutical company. There was concern that engineers were taking an excessive amount of time to deal with machine breakdown. Initial focus for the HTA was on the tasks and skills of maintenance engineers, but these proved difficult to understand.

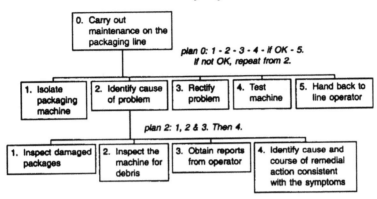

Figure 2.4   Higher levels of redescription of a maintenance task carried out by packaging line engineers.

When a broader perspective was considered, namely an analysis of the general task of maintaining production in the packaging line, the activities of line operators were also taken into account and the problem became tractable.

Figure 2.4 shows the top levels of the HTA of the engineer's task. This shows a reasonably standard set of fault diagnosis routines. Figure 2.5 shows the line operator's task. This analysis was taken from the perspective of the system to establish the set of things that needed to be done to operate the machinery. The key here is how the operator carries out operation 3.3 'Deal with system trips and major blockages'. To examine this, the HTA was taken from the operator's perspective. Figure 2.6 shows what operators actually did when problems arose. Because the maintenance engineer was usually on another job when they called, line operators tried to use their time usefully and cleaned up their machines so that they would not be delayed for a restart. The problem with this strategy was that they destroyed all the information the engineer needed to resolve the problem. Because the two jobs were treated separately, the operators never realised the

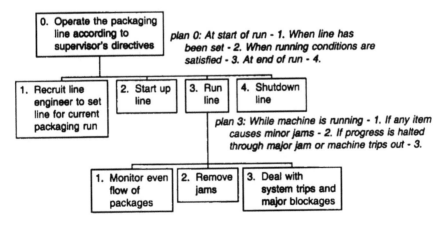

Figure 2.5   The packaging line operator's task.

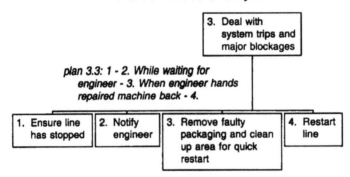

Figure 2.6   *How the line operator initially dealt with faults on the line.*

consequences of their actions and it never occurred to the maintenance staff that their key information was being destroyed in this way. Figure 2.7, represented a more effective way for the line operators to work.

There are numerous other examples of the analyst becoming too focused too soon. Some examples involved more than one person, as with the packaging operation, and other examples involved other activities that the operator was engaged in which had not yet been considered in the analysis. In one such example, profitability in a plant depended on operators working out how to correct for a formulation in a product whenever it was off-specification. This was a difficult task and a serious problem because it occurred so frequently. By analysing the wider task, it became clear that the problem only arose when an earlier cleaning operation was not done properly. The reformulation problem mainly arose because the vessel was contaminated by previous batches. Improving the clean-out procedure was straightforward in comparison with training operators how to calculate the reformulation required.

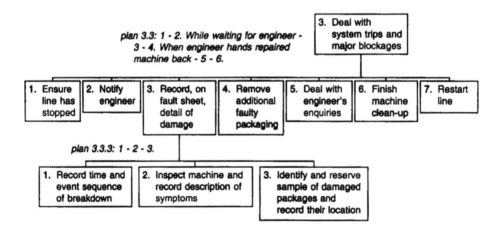

Figure 2.7   *How the line operator should deal with faults on the line to provide the engineer with better support.*

2. Explore its constraints

There are constraints on how the operator makes responses and there are constraints on the options that the analyst can choose in making recommendations.

There are many *operating constraints* that affect performance. Poor environmental conditions can directly affect perception. If illumination is poor, the operator may not be able to discriminate crucial signals. If there is a lot of background noise, sound cues will be masked. Poor discrimination will cause signals to be missed or information to be misperceived and, therefore, mistakes made. Where conditions are cold and draughty, muscles can be affected. Where circumstances are such that errors are critical, stress can affect decision-making. If the analyst can appreciate the factors that can influence performance, then this knowledge can be brought to the task analysis as a means of better understanding the problems with the task.

Design constraints are those factors which limit the design options that can be pursued. For example, if people carry out a task infrequently, training benefits may be limited and job-aids preferred. If there are strict cash limits on what can be expended in improving task performance, the analyst must be aware of these. If there are constraints on who can be employed, then this too will constrain the analyst from considering some personnel options. Some constraints are technical. Thus, electronic devices causing sparks cannot be used in inflammable areas - so, certain sorts of tools of job-aids cannot be prescribed.

The analyst may also become aware of factors which are not strictly rational, yet still need to be observed. If there is a culture in which people do not use job-aids, then this option might best be avoided if possible.

Exploring constraints tends to be informal and intuitive and either kept in the back of the analyst's mind or recorded in a separate note. Sometimes constraints are not obvious at a particular level of description and are only recognised when the goal is explored further.

3. Judge whether the goal will be met to an acceptable standard given prevailing circumstances

Earlier in this Chapter emphasis was placed on the importance of the analyst making a rational decision concerning whether an operation or goal should be examined. It was stressed that task analysis is a practical activity and should, therefore, be bound by the same constraints as other managerial and design activities. Emphasis was placed on the P x C rule to ensure that proper account is taken of both probability of inadequate performance and the costs of inadequate performance in making these judgements.

4. Cease further analysis at this point

If, following Stage 3, the value of P x C is considered acceptable - that is, the risk of leaving things as they are is acceptable - then analysis of that particular goal can stop. As the analyst delves deeper into the task, this outcome will arise more and more frequently until all of the task is dealt with in sufficient detail.

5.   Examine the operation and the human-task interaction

Where performance warrants attention, examination should first consider the *operation* and the *human-task interaction* with a view to generating a *design hypothesis* that will overcome the performance deficiency or help make a judgement concerning the cause of this weakness in performance. This phase is central to HTA. It may involve attempting to understand information processing, cognition, attitudes, etc. or it may be done informally by the analyst relying on experience or human factors knowledge. This issue of dealing with design hypotheses was discussed more fully earlier.

6.   Estimate the cost-benefits of the hypotheses

One common outcome which follows the examination of a human-task interaction is that the analyst proposes a means by which performance could be improved. There are often many ways of improving problems associated with the performance of tasks, with some more expensive than others. An important aspect of task analysis, then, is *cost-benefit analysis* applied to the hypotheses under consideration. If the costs of an innovation exceed its benefits, then that innovation is not worth pursuing. Equally, if several possible, equally valid, hypotheses are considered, then the least expensive should probably be preferred. Again, this is a judgement that is made as a routine within the cycle of activities in task analysis. It cannot be a formal judgement at this stage, but formal cost benefit analysis may be required later on in order to secure funds for a project.[19]

Cost-benefit analysis becomes increasingly important as technologies for controlling systems or dealing with human factors solutions become more expensive. Training simulation is a good illustration because the technology required to obtain high fidelity simulation in many domains is very expensive and may not appear to be justified, even in terms of the costly events it may avert. It is, however, important to recognise that as task analysis progresses and further opportunities to use the same equipment or technology present themselves, benefits may start to overhaul costs. Therefore, it is important that no potentially useful hypotheses are totally discarded on cost grounds, but kept alive to be reviewed later.

It will be noted that issues of cost-benefit analysis are related to issues of stopping analysis with regard to the P x C rule. A proper cost-benefit analysis would compare regimes and not simply deal with the costs and benefits of design innovations in isolation. Calculating the 'cost-benefit' factor for any innovation entails costing the innovation fully and identifying benefits broadly. Costs include capital and recurrent costs associated with the innovation and the costs that will be incurred as a consequence of risks that will still prevail. Any innovation will merely serve to reduce risks; eliminating risk entirely is fanciful. Moreover, benefits can include improved productivity, but they will also include hidden benefits such as the additional expertise the company has now gained as a result of the innovation. Introducing virtual reality, for example, may not be justified in terms of the cost-benefits of a single project but may make sense by taking a longer term perspective.

7. Record the hypothesis and cease further analysis at this point

If a hypothesis is judged acceptable, then the analyst can cease at that point. The analyst should record the hypothesis and then move on.

8. Attempt to redescribe the goal

Where the analyst has been unable to generate a suitable hypothesis, the redescription of the goal into its subgoals and their plan is warranted. This is the hierarchical feature that characterises HTA and will be discussed extensively later on.

9. Seek advice or review constraints

Sometimes redescription proves impossible for the analyst. This may be because no way of redescribing can be seen by the analyst or it may mean that no way can be seen within the given set of resource constraints. To resolve the problem the analyst may seek advice. Such advice may provide help in suggesting a method of redescription, or it may provide a more acute examination of the human-task interaction, leading to a design hypothesis. Redescribing operations or goals is a skill that develops with experience.

10. Identify the next goal to be examined

Once the HTA is underway and a first redescription has been completed, the analyst must choose which of the remaining subgoals should now be given attention. Strictly speaking it does not matter which remaining subgoal is dealt with next. In practice it can help to work forward in time in order to be better acquainted with what has already happened in the task.

Using the framework

The cycle of activity described in Figure 2.3 can be used to analyse all tasks. In doing this, a number of judgements must be made if the analysis and the recommendations are to be sensitive to the requirements of the client. It is also worth noting that this process results in the analyst recording a hierarchy of goals, subgoals and plans.[20]

Concluding comments

This chapter has set out the processes of analysing a task in a systematic way. Any task analysis project entails the analyst working with people who provide information about how the task is done, why it is done, who is involved in the task, the context in which the task is done and the constraints that must be observed when seeking alternative ways of doing things.

Three main considerations were listed to provide a useful basis for any task analysis project. First, the analyst must judge which parts of the task warrant investigation. This is crucial to ensure that time and other resources expended during the task analysis project are used to best advantage. Elements of tasks where performance is non-critical or where performance is good warrant less attention that those areas where the consequences of human error are costly and where people are likely to make mistakes.

Second, the analyst must be able to judge, for critical elements of the task, what factors give rise to concerns over operator performance and what steps can be taken to eradicate these problems. This involves analytical experience and use of various methods to help the analyst gain further insight. It is here, especially, that other analytical techniques may be brought into play to supplement the analysts own knowledge and experience of tasks.

Third, the analyst must know how to redescribe operations into greater detail. In HTA, this is done through redescription into subgoals and their plan which governs the conditions when subgoals must be carried out to meet the overall objective.

These basic ideas are accommodated within a framework for task analysis. By following this general procedure and recording the outcome, the characteristic HTA task hierarchy emerges. Thus, HTA is a strategy for analysing tasks and not simply a method of representing tasks.

This task analysis framework provides a systematic strategy for the analyst in which various analytical skills and, sometimes, other analytical techniques are brought into play. Task analysis, at a practical level, cannot be conducted in a rigid procedural manner. With only a little experience, the analyst is able to work through these considerations routinely. The framework serves to outline the underlying approach and show how the various considerations need to be made systematically.

The next chapter will focus attention on the processes of developing plans in HTA. This is an important aspect of HTA since plans are the key to accounting for complexity in tasks.

Chapter 3

Plans and complexity

Plans are an essential element of describing tasks using HTA, because they represent the conditions when subgoals are to be carried out to satisfy the goal being redescribed.

This chapter sets out most of the commonly occurring elements that account for plans identified in HTA. They include plans which describe various forms of fixed procedure, plans concerned with decision-making and plans which operate in cycles of activity.

When different plans are combined together, they can account for considerable complexity in tasks. A useful consequence of this is that complex tasks often be better understood by representing them as a hierarchy of simpler plans. Strategies for unravelling complex plan in this way are illustrated.

Introduction

Plans in HTA indicate how subordinate operations are organised in order to meet their common goal. Plans comprise a limited number of *timing and sequencing* relationships. For example, in some plans suboperations are carried out in sequence, while in others they may need to be carried out together. In other plans operations are contingent upon certain system events occurring. These timing and sequencing elements may be combined in different ways in actual tasks to account for different sorts of operator performance. Plans also combine to create apparently *complex* performance - the apparent complexities of human performance can often be accounted for by describing simple plans in combination with one another.

There are a number of ways in which the *context* in which an activity takes place affects the performance of the task. There are many plans in which carrying out an operation is contingent upon system events and not simply on which operation was carried out previously. Frequent events provide opportunity for practice and consolidation of skill. Therefore, different strategies to maintain performance may be adopted in comparison to operations which are required only infrequently. Thus, an important aspect of task analysis concerns understanding how events affect plans.

The way in which plans combine to account for greater complexity can also be exploited as the basis for analysing apparently intractable complex plans. When confronted by an apparently intractable plan during task analysis we can often make sense of the complexity by seeking intermediate goals and their associated hierarchy of simpler plans.

Different sorts of plan

When we carry out HTA we often recognise similarities between plans, even when these arise in different contexts. It is helpful to recognise how different sorts of task activity in unfamiliar contexts can be described in ways that we already understand. The following six features are prominent in plans:

- Fixed sequences

- Contingent sequences

- Choices

- Optional completion

- Concurrent operations

- Cycles

Fixed sequences

A fixed-sequence component in a plan is where a specified second operation is carried when a first goal has been successfully attained. For example, to use a toaster a slice of bread is inserted then the lever is pressed down and so on. Assembling a recently purchased personal computer entails plugging in the keyboard, then plugging in the mouse, and so on. To save text in a wordprocessor, the user may need to move the cursor to the 'File menu', then hold the mouse button down to 'pull down' the file menu, move the highlight down to the word 'save', then release the mouse button. This is a fixed procedure and can be represented as in Figure 3.1. The behaviour necessary to conduct a small task such as this includes the capability to carry out each of the suboperations and the ability to do them in the right sequence. The *cue* to carry out the second operation is the *feedback* that shows that the first goal has been attained.

Knowing what to do next in any procedure can stem from a number of different sorts of behaviour. One option is for the operator to learn and remember a sequence of subordinate operations. Another option is for the operator to follow a written instruction. In this case the operator still needs to know *how* to do each of the suboperations, but the written procedure provides the means of prompting the plan. Another option is for the operator to work out what to do next. Computer interfaces are designed to prompt the user by appropriate choice of names for menus and commands. Thus, the user who wishes to save a piece of work may not recall what to do but is prompted by seeing the word 'file' at the top of the screen. This may cue an action to pull down the menu, when the next clue is revealed - the menu option 'save'. A familiar action such as 'save' may well be remembered, but something less familiar, such as formatting tables in text may not be remembered, but may still be achieved by sensible guesses concerning where commands reside. So, with trial and error and then practice, additional routines become available. Thus, the 'plan' in Figure 3.1 sets out the *criterion* for performance in operational terms. It does not imply that a specific kind of behaviour is always necessary to guarantee success.

Contingent sequences

In tasks where the operator interacts with a system with its own dynamics, completion of the previous action is rarely suitable as the cue for prompting the next action. A

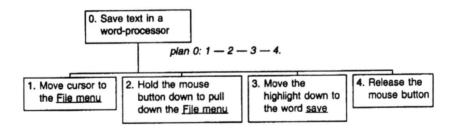

Figure 3.1   Fixed sequence element in a plan. Each step is carried out following completion of the previous step.

simple and familiar example is cooking. The rate at which someone works through a cooking task is contingent upon prior stages being reached and not simply a consequence of actions that have been completed. If we are boiling an egg in a pan then: we place the egg in the pan; cover it with cold water; place it on the stove; turn the stove on; watch the egg until the water boils; adjust the cooker to maintain the water simmering; then start a timer; monitor the timer; then when 3 minutes has elapsed remove the egg from the water. In this example, the operations have been carried out in sequence, but the cue for the next action is not always the feedback that the previous action is completed successfully. Thus, we start the timer at a stage when the water has just started to boil. The time to boiling from cold cannot be predicted with certainty as it is affected by the size and temperature of the egg, the initial temperature of the water and the thermal equivalent of the pan. All the cook can do is monitor visually that the water has started boiling. Figure 3.2 shows the simple task analysis of boiling an egg. There are fixed sequence elements in this plan, but the step between turning the stove on and starting the timing is a *contingent sequence* element.

Contingent sequence elements are commonplace and critical when examining tasks concerned with controlling complex systems. Typically, a first operation will cause a change to operating conditions that then cause system parameters to change. Process control situations, such as operating a petroleum refinery or power generation plant, provide obvious examples. Figure 3.3 shows a simple example of a plan which involves a contingent sequence taken from a task in which the operator must bring a vessel to a required operating temperature. An action is carried out to introduce steam to a vessel (here measured in psi - pounds per square inch), then, after 2 minutes a drain valve has to be closed. Because physical objects take time to heat up the operator cannot assume immediately that the desired effect will have taken place, so the operator must monitor the temperature rise. When the temperature reaches 100 degrees, the steam is turned off so that the energy still in the system will cause the temperature to reach the target without overshooting.

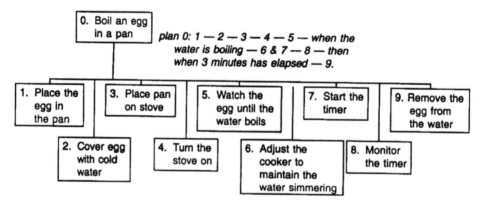

Figure 3.2   A contingent fixed sequence element is where a subsequent operation is cued by an event other than the satisfactory feedback from the previous operation, i.e. the cue for operations 6 & 7.

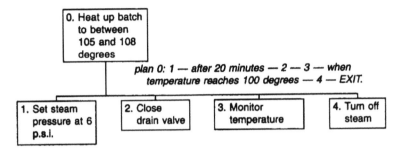

Figure 3.3   Typical contingent sequence plan elements containing two contingencies - one concerned with time and the other with the state of the system (temperature).

The example given is that of a physical system. Contingent sequences are common in physical systems because of inertia, because of complex linkages between components, because of the presence of automation and because of varying physical conditions. These factors cause simple input actions to be altered in ways such that their consequences are either delayed or combined with other actions. The same occurs in organisations where people collaborate in processing information. As information flows through an organisation, it can be delayed, because meetings are delayed or because staff are overworked or uncertain about decisions. Information is combined with other information to disguise its source. Information is affected by environmental factors and how other members of staff perceive these factors. In both physical and social systems where people must collaborate subsequent action may need to wait until colleagues have completed their tasks. Thus the cue for an operator to do the next thing may be a message from a colleague or satisfactory feedback from the colleague's action. This may appear to be quite detached from what the operator did previously.

There are many occasions when the analyst may choose whether to describe a plan in terms of a fixed sequence or a contingent sequence. For example, in the case of boiling an egg, the analyst could describe the task as an operation as in Figure 3.2 in which plan 0 contains a contingency, whereas in Figure 3.4 plan 0 contains generally fixed sequence elements. Here, the contingency is handled by operation 4. It does not matter which is chosen, provided that the plan, taken in conjunction with its associated operations, represents the task in the way that the analyst intends.

The importance of distinguishing between *fixed sequence* and *contingent sequence* plan components is to remind the analyst that the cues for the next action sometimes rest with the action feedback from the previous action and sometimes depend upon another system cue.

Choices

There are many situations where different actions must be followed according to different sets of circumstances. This arises, for example, in maintenance tasks where a fault-finding operation determines which remedial action should be followed. Equally, medical

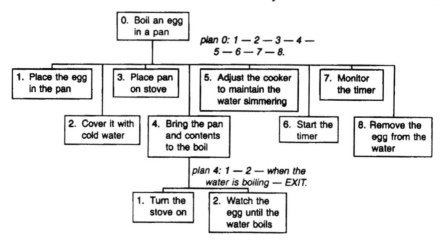

Figure 3.4   This represents the same activities as Figure 3.2 but plan is represented as a fixed sequence.

tasks require different treatments or care regimes according to the condition of the patient. A sales person will proceed with customers differently according to what has been found out about the customer's requirements. Just as recognising a contingent sequence prompts the analyst to record a previous *monitoring* operation, so a choice in a plan indicates that a *decision-making* operation is first undertaken. Decision-making entails obtaining information about the choices to be made.

There are different sorts of skill concerned with decision-making operations. Decision-making may entail making discriminations such as distinguishing between a red and a green light to decide how to progress with a task. Distinguishing between red and green for people without sight defects is usually straightforward provided illumination levels are satisfactory and the task is not demanding attention elsewhere. So, identifying this sort of decision task could signal the need to screen job applicants for eyesight, it could stress the importance of providing adequate illumination, it might prompt a reappraisal of how jobs are designed or it might remind the designer to provide alternative signals for making a decision which was not so reliant upon the capabilities of the operator.

Other situations will be less clear-cut. For example, a sales person may be required to distinguish between customers who will respond to a sales pitch and those who will be repelled by such an approach. If this is done well it would save time and improve the sales success rate. Establishing this decision-making skill will depend upon training and experience. A surgeon may need to judge which patients will respond to surgery and which ones will not. This will entail making clinical tests, reviewing patient records and making risk judgements from this combined information. Equally, a technician may need to carry out fault-finding to determine what needs to be done to repair an item of equipment. All of these types of decision-making are crucial to effective industrial, commercial, social, medical, administrative and recreational activity. To make choices and decisions is often the reason why people are employed.

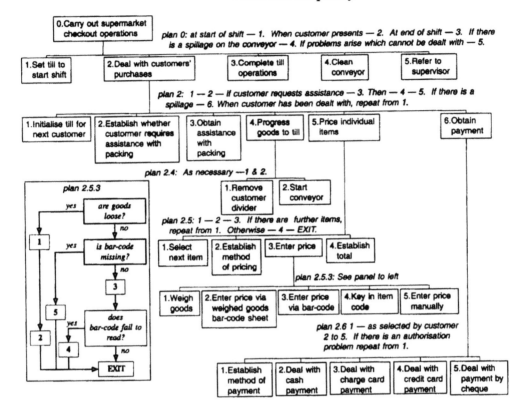

Figure 3.5   A supermarket checkout task. Plans 2, 2.5, 2.5.3 and 2.6 all illustrate various aspects of handling choice and decision-making.

Choice and decision-making can be represented in a variety of different ways - there are many examples in the task analyses represented throughout this book. Figure 3.5 shows the analysis of a supermarket check-out task, which contains a number of different choices within plans represented in different ways. The check-out operator waits for the next customer, then processes items from the customer's trolley then establishes the charge to be made and obtains appropriate payment.

In 'deal with customers' purchases' (operation 2) one responsibility is to ensure that customers obtain assistance with their packing if they so desire. Thus, plan 2 contains a choice element where a decision is taken concerning the customer's requirements. In 'obtain payment' (operation 2.6) the operator must establish how the customer wishes to pay - presumably by asking them. The alternatives are set out for the operator to follow according to the customer's choice.

In 'enter price' (operation 2.5.3) the operator must determine how each item will be dealt with. In Figure 3.5, plan 2.5.3 has been set out as a decision flow diagram. It could equally have been set out as a set of rules as follows.

If goods are loose - 1 - 2.

If goods are not loose, but the bar-code is missing - 5.

If goods are not loose and bar-code is available - 3.

If bar-code fails to register - 4.

Optional completion

Many tasks require that the operator conducts a certain number of operations but without constraint on the order in which they are carried out. These tasks include maintenance tasks and preparatory tasks.

In preparing an aircraft for flight, for example, certain subsystems must be shown to be operational before takeoff. Sometimes it does not matter in which order these things are done - the main requirement is that they are *all* done. One obvious solution to support performance is to provide a 'checklist'. Another common practice is to turn the plan into a fixed procedure - while the order in which actions are carried out may not

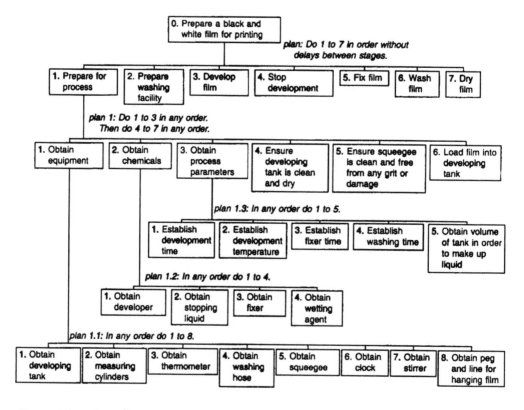

Figure 3.6   Part of the task analysis of developing a black and white film. This shows why the satisfactory completion of preparatory tasks sets the context for later activity.

matter from an operational perspective, it can often be helpful to make the operator do things step by step so that nothing is forgotten. Another advantage of prescribing a procedure, rather than allowing the operator too much freedom, is that tasks can be organised to be done in a convenient manner. If some activities require safety equipment to be used, then these might sensibly be grouped together. A further reason is recoverability. If some faulty items are detected soon enough, there may be time to rectify the problem before its consequences become critical.

Preparatory tasks are where items of equipment and materials must be assembled prior to lengthy or critical operations taking place. Figure 3.6 shows how an optional completion task could be represented. The example is that of making preparation for developing a black and white photographic film and concentrates on basic preparations being made to equipment and preparing the film for processing. There are numerous bits of equipment that must be assembled and prepared and various chemicals that must be obtained and prepared. Once the process starts it must progress with haste to ensure that chemicals are not left to get cold and the film is not left too long in the different processing stages. The stages of preparation represented in Figure 3.6 ensure that, once started, there is nothing to impede the film being processed. So while the preparatory operations seem easy to accomplish, the consequences of not doing them properly and to time has serious implications for other parts of the task. The importance of these influences on industrial and commercial processes is well understood.

Concurrent operations

Often, several operations must be carried out together to meet the requirements of a task. At first glance, these are easy to describe - the plan merely states that two or more operations must be carried out at the same time. However, this can imply a number of different things.

Time saving

Some tasks are required to be carried out at the same time simply to save time. Where industrial and commercial processes do not demand complete attention from the operator, there are opportunities to get other things done. If a clerical officer is awaiting a decision from a supervisor, there may be time to get some filing done. If a process operator is waiting for a vessel to warm-up, there may be time to start to prepare materials for the next batch. If a maintenance technician is waiting for a new part to arrive, then there may be time to start looking at another repair job.

Coordination

Some tasks require operations to be carried out together in order to take account of the progress being made in each of them. Several examples are associated with *skilled* performance, where the operator must carry out operations concurrently in order to coordinate different parts of the activity.

A good example is that of carrying out a hill start in a manually controlled car. The driver must coordinate use of the handbrake with appropriate control of both clutch and accelerator pedals. The driver must release the clutch whilst increasing the accelerator until the car can be held stationary on the hill. Then the handbrake is released and the car can move forward. Releasing the clutch too far without sufficient acceleration will cause the car to stall. Too much acceleration will cause the car to leap forward. The driver's skill depends upon relating the affects of depressing the accelerator to the release of the clutch. It is a skill that comes about through practice. Plans in HTA are not helpful in specifying this sort of task coordination. Coordination of these elements is a skill that the driver will acquire through appropriate practice and the method adopted by a particular driver may be different from that adopted by someone else. So, while we may say that the driver plans and coordinates these various suboperations, we cannot state a clear plan to serve our HTA. The analyst would have to cease redescription at the goal of carrying out a hill-start, make a note that a skill must be acquired, then specify appropriate training to enable the skill to develop. To this end, the analyst could list what he or she believes are the constituent operations - use of clutch, accelerator, handbrake, engine noise - and then specify a schedule of practice through which the task can be mastered. This is often sufficient in practical task analysis projects, although the learner driver may have learned to rely on other cues that the analyst may have overlooked, for example, engine vibration.

Whereas the driving skill discussed might be referred to as *cognitive* coordination, there are many other concurrent task elements that can be described as *operational* coordination. There are many instances where two operations must be started together in order to be finished together to be ready for a third operation to commence. This is illustrated in Figure 3.7 where Unit 3 depends upon Units 1 and 2 being ready. If Unit 1 takes 3.5 hours to prepare and Unit 2 takes 2.5 hours to prepare, then the respective starting times of each unit needs to be worked out. Figure 3.7 is a very effective way to represent this sort of plan. Typically in industrial processes timing is crucial. If Unit 1 was ready before Unit 2 was ready, then this would cause wasted energy in maintaining Unit 1 in a holding state longer that is strictly necessary. Also intermediate product in Unit 1 could spoil. Concurrent operations such as these are not carried out simply to save time, but also to avoid costs such as energy, labour and safety risk. If staffing levels need to be increased to accomplish this, then this is a cost that may well be justified.

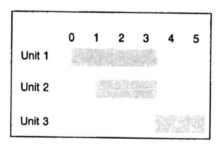

Figure 3.7   A time-line plan for starting up 3 interdependent units.

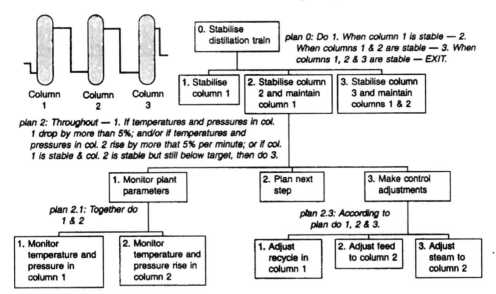

Figure 3.8   Controlling a distillation train. Columns 1, 2 & 3 must be started and stabilised in turn. As columns are started, those already running must be adjusted as feed taken from them will de-stabilise their condition.

A second form of operational coordination is where different process parameters need to be controlled in conjunction with one another. This is very common in starting up complex processes. Figure 3.8 shows three connected distillation columns and the HTA describing part of their start-up procedure. Distillation is a common industrial process concerned with purifying liquids by separating them according to their different boiling points. Distillation is the process used in making whisky, taking advantage of the fact that water and alcohol boil at different temperatures. A distillation column is a tower into which a liquid feed is pumped - for example, crude petroleum in the case of oil refining - and then heated to a temperature in which one liquid vaporises while the remainder stays in its liquid form. Sometimes it is required that further distillation takes place to further purify one of the liquids from the top or the bottom of the first column. A chain of such columns is a distillation train.

In the task in Figure 3.8 the operator first stabilises column 1 by adjusting operating conditions to ensure that the levels, flows, temperatures and pressures throughout the column are in accordance with criteria specified by the plant engineer - the procedures for doing this would be set out in a redescription of subgoal 1. When these process parameters are established, the product can be fed forward to column 2. As soon as this 'feed-forward' commences, column 1 is destabilised. Therefore, while the operator tries to stabilise column 2, attention must also be paid to column 1 to bring it back to its own stability. This involves time-sharing between monitoring the effects of actions on column 1 and on column 2 as stated in plan 2.1. Further aspects of concurrent activity relate to the goals which follow monitoring - 'plan next step' and 'make control adjustments'. These must also take account of both columns, although this is not time sharing but a more complex diagnosis and planning task in which the implications for both columns 1 and 2 must be taken into account. These activities are further complicated when column 3 is introduced.

Cycles

Many tasks require the operator to repeat an activity until a particular condition has been met. This applies in conditions where it is required that a service is provided continually over the period of a shift. This used to be common in assembly tasks in industry where operators were required to repeat a limited number of assembly operations for the duration of their shift. These tasks can still be found but are less common now because of changes to manufacturing methods. Another example was the supermarket check-out operation, illustrated in Figure 3.5. Plan 2 in that analysis was concerned with the cycle of activity involved in dealing with a customer - from initialising the till, through to obtaining payment. Plan 2.5 was another cycle concerned with how the operator priced an individual item. This cycle was part of the first cycle, so the observed behaviour could appear quite complex.

Procedural cycles

Dealing with customers in any service industry entails cycles of activity. Figure 3.9 shows the analysis of part of a typical tele-sales procedure. If there are no customers, the operator will monitor a screen to respond when a customer calls. When there is a

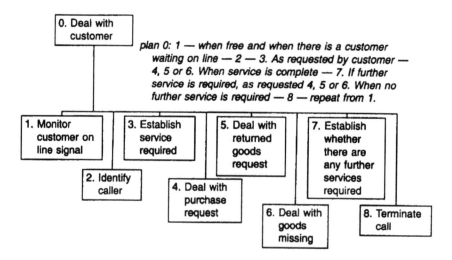

Figure 3.9   Deal with a tele-sales customer. This cycle of activity will be repeated as long as the operator is on shift. Each of the other operations would be further redescribed, adding to the complexity of the task.

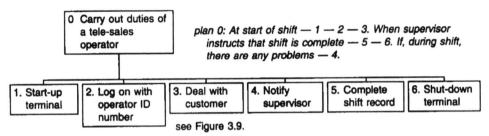

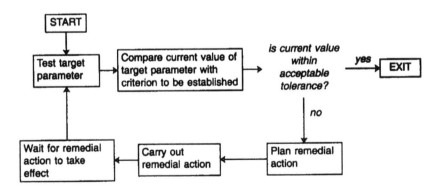

Figure 3.10   Typically, procedural cycles fit into jobs as a component of a general shift cycle. The main activity is repeated until terminated by an instruction from the supervisor or by the clock.

call, a common cycle of activities is followed, from identifying the caller and the service required through to arranging delivery and payment. On termination of the call, the operator returns to monitor the next call.

The cycles of activity just described can be called *procedural cycles* because they link the end of a procedure back to its start. Generally such procedural cycles continue for as long as specified. If we were to analyse the overall job of a tele-sales operator it would probably look something like Figure 3.10. Typically in the examination of such jobs, we see a shift pattern where the operator logs on at the start and logs off at the end together with completing reports to management. In between are various activities, including the stage of dealing with customers, as in Figure 3.9.

Remedial cycles

Procedural cycles can be distinguished from *remedial cycles*. A remedial cycle is where the operator is engaged in a set of activities to achieve a required goal. Typically,

Figure 3.11   The common remedial cycle is where the operator executes a cycle of testing, comparing, adjusting, and waiting, until a criterion has been reached.

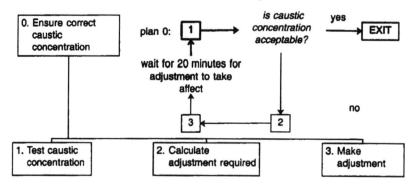

Figure 3.12   The common remedial cycle in a task analysis.

the operator will make a test of a parameter, calculate the deviation from target, plan a method of recovery and implement that method of recovery. Then, when the remedial action has had chance to take effect, the operator will re-test. This is a common control structure and has already been alluded to several times in this book. It can be represented diagrammatically as in Figure 3.11.

The remedial cycle is one of the most useful plan structures encountered in HTA. It is found in all control tasks, including process control tasks, clinical tasks, training, supervision and other personnel tasks. In clinical tasks, a doctor may need to make adjustments to drug doses in order to achieve an appropriate level of stability in the patient. This will require administering a drug, allowing it to take effect then calculating

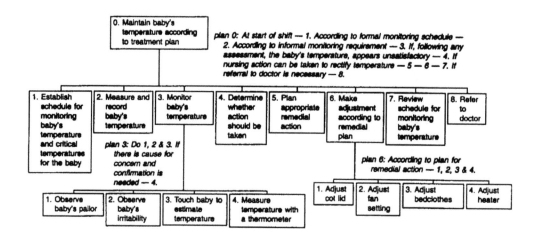

Figure 3.13   Maintaining a baby's temperature in hospital — an illustration of a remedial task in which the cycle of activity is implicit within the plan.

any adjustments required. Process operators are also often required to adjust formulation to meet target specification. This is done by a cycle of testing product, planning change, adjusting formulation, waiting for the effect to take hold and readjusting as necessary. An illustration of a typical remedial cycle is shown in Figure 3.12. This illustration concerns correcting the amount of caustic soda in a material being processed in a small plant. The redescription has three subordinate goals that form part of the cycle. The plan sets out the conditions for these operations. Crucial within this plan is the need to wait for 20 minutes between making adjustment and retesting. This allows the change to take effect. Similar plans account for activities concerned with correcting formulation and with adjusting temperatures.

A more complex example is the task of maintaining the temperature of a baby in an intensive care unit, shown in Figure 3.13. Nursing staff are employed to maintain the baby's comfort and continually monitor the baby's vital signs to determine how he or she is responding to care and treatment. One of the key vital signs to monitor is temperature. The nurse estimates temperature from observation and through touch, as well as taking the baby's temperature with a thermometer. Judging whether the temperature is within an acceptable range is more than simply comparing it with the average for a healthy baby. Consideration must also be given to the baby's current circumstances and current treatment. Planning to correct temperature must also take account of the baby's circumstances. A baby in hospital being nursed is at risk and so the nurse will need to check the consequences of any action by frequent monitoring of the baby's comfort. A baby whose health is at risk will need to be attended to more frequently and more urgently than one who is nearly ready to go home. So, in this example, it is inappropriate to specify categorically how long the nurse must wait before next looking at the temperature. The nurse and other care staff must be more flexible and judge for themselves the frequency of monitoring as part of their planning.

With respect to the top level of redescription, operation 1 is concerned with the nurse establishing the criteria by which the baby's temperature should be managed. Normally there is a standard schedule for recording vital signs, but the frequency of measurement might be increased by a doctor prescribing a treatment plan. This schedule specifies the frequency of doing operation 2. In addition to this formal measurement, the nurse will routinely glance at each baby to determine whether things are alright. Thus, 'monitor baby's temperature' (operation 3) may be done informally according to the nurse's skills in observation and judgement. If the nurse was particularly concerned, then a thermometer could be used, although skilled judgement is less intrusive for the baby. The plan states that operation 3 is carried out 'according to informal monitoring requirement'. This means that the nurse will do it intermittently or, if there is any reason to be concerned - from information from previous shifts or from direct experience - the frequency of this monitoring may be increased. If the temperature is judged unsatisfactory, the nurse should first reflect on whether any action should be taken. The condition of the baby or the baby's immediate past history may suggest that this deviation is acceptable. If the deviation is not acceptable, then action is necessary. There may be conditions where the nurse is instructed to refer problems directly to the doctor. Alternatively, the nurse may have discretion to act directly. Operation 5 entails deciding which things should be done. These things are then done in operation 6. Then operation

7 stipulates that the schedule for monitoring the baby in the near future needs to be reviewed. Thus, greater attention will be paid to a baby causing concern. Operation 7 implies an informal judgement is made which will guide the nurse's subsequent actions in prompting when operation 3 is next done. This judgement is also the sort of thing that is recorded and communicated to colleagues, especially at the next shift change.

The outcome of plan 0 is a cycle of activity, which follows the general patterns set out in Figure 3.11, but which is not stated as explicitly as the example in Figure 3.12. In the case of the nursing example in Figure 3.13 the schedule for monitoring and waiting is determined by the operator and then used within the plan to guide other actions. Clearly, when observing actual performance, nurses are not explicitly doing things in this way, but examination of their task shows that these things need to be done and, therefore, these are the skills and decisions that need to be learned.

Composite plans

Despite identifying these common types of plan element, most plans in HTA are composites. Indeed, it is rare to encounter plans that are exclusively one type rather than another. Analysts must use plans to represent the task and should not feel compelled to fit them into tight categories. Most important is that the analyst scrupulously follows the rules of redescription in making sure that plans provide a proper account of the task.

Unravelling complex plans

Sometimes plans appear intractable. They seem too complex to enable sense to be made of them using such explicit procedural devices as plans. This chapter will conclude with an illustration of how such issues might be explored by working out how a single, apparently intractable, complex plan may be better represented as a hierarchy of simpler plans.

Operators in the industrial complex illustrated in Figure 3.14 were responsible for managing the production and consumption of chlorine gas. Production of chlorine takes place through a process of electrolysis of brine. A hundred or so brine cells are contained within each of three cell rooms. Electricity is required to enable this electrolysis process. In addition to chlorine gas, a number of unwanted by-products are produced. This gas mixture is pumped to the gas handling plant where it is purified. The purified gas is then sent on to be used in other plants on the site. One use is in the liquefaction system where chlorine gas is condensed into a liquid under pressure and bottled. In other cases the gas is sent to user plants where it is used to create other products for sale. There are three cell rooms on the production side. On the consumption side there are two user plants and one liquefaction unit with a tail gas system which looks after by-products of liquefaction. There is also an emergency absorption tower, a necessary precaution to cope with excess gases being produced which cannot readily be used.

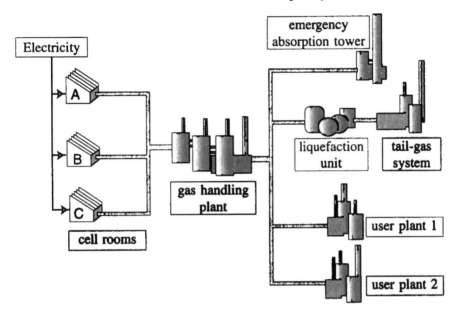

Figure 3.14   The chlorine production complex.

The person operating this system is called the *chlorine controller*. The chlorine controller's job is to optimise the production of gas and distribute it according to the production targets of the various user plants. For this task, electricity was more expensive in the daytime than at night and the chlorine controller had to take this into account. Also, although they share a similar technology, the cell rooms A, B and C operate at different efficiencies. If one cell room, say B, is shut down for maintenance, then production has to be done via A and C. However, if B is the most efficient producer of chlorine and then comes back on-line, the controller has to judge whether it is better to switch the electricity from A or C to B, bearing in mind that the change-over process will waste electricity anyway. But if B becomes available just before a tariff change when production must be cut back anyway, then a change may not be worth making.

Even if further chlorine could be made, there must also be sufficient capacity in the gas handling plant to treat this extra gas. The gas being made will normally be · distributed between user plants 1 and 2 in order to meet certain production targets. Sometimes liquefied gas will be required. Moreover, liquefied gas may be stored either to feed user plants or to sell to other companies. The emergency absorption tower is an automatic safety facility and not used as part of normal resource balancing.

Just as cell rooms may be shut down for planned or unscheduled maintenance, so too are user plants. If there is insufficient capacity in liquefaction or no more capacity in the other user plant then the current rate of production must be cut back to avoid the risks of excess chlorine in the system.

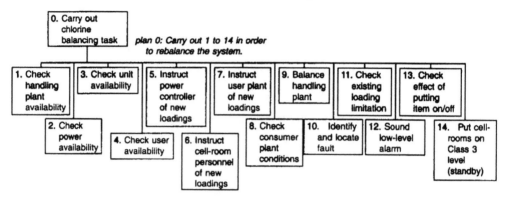

Figure 3.15   A first attempt at redescription of the chlorine balancing task.

This was an important planning task that required skilled execution. It affected the profitability of the company and the safety of personnel and the environment. As with many tasks, operating skill was seen as a black art. Questioning managers and operatives about how this task was carried out enabled a list of fourteen common actions to be drawn up. But little insight was offered concerning how these actions were organised. The initial task analysis is shown in Figure 3.15. It was generally agreed that each of these activities was reasonably straightforward for operators to learn and to do. The plan in Figure 3.15 merely reiterates its goal and is generally unhelpful. The purpose of the task analysis was to find out how this plan could be expressed more informatively.

Given that plans are driven by events, a strategy for examining this unspecified fourteen operation plan was to identify a range of events to which the chlorine controller was likely to have to respond. Examination of the system in Figure 3.14 suggested the major categories of events that would occur. Events could be related to electrical *power* becoming available or being withdrawn; *cell rooms* becoming available or being withdrawn; the *gas handling plant* becoming available or being withdrawn; or the *user plants* becoming available or being withdrawn. Moreover, the need for these changes could arise through long term planning or on a shorter time-scale. Long-term planning changes included changes due to the tariff change when electricity became more or less expensive. Long-term maintenance changes enabled warning to be given when units were to be taken out of service. Production schedules would change as market demand changed. Shorter-term changes included power becoming available because other plants on the site were temporarily off-line, or user plants warning that they would be going off-line shortly to undertake maintenance. As user plants lost production, so their production schedules would be affected. When they were again available for production, the controller had to take steps to make up the deficit. Emergency situations were not foreseen. They included equipment breakdown, gas escapes and so on. This distinction between the different degrees of warning were generally clear-cut.

To analyse the task, chlorine controllers were presented with a number of events that represented the main categories of disturbance to the system and asked how they would deal with them using the 14 operations in Figure 3.15. The sequences that they

Table 3.1   Clusterings of operations in the chlorine balancing task.

Reviewing resources	Dealing with faults	Alerting	Changing the system
1 Check handling plant availability	10 Identify and locate fault	12 Sound low-level alarm	5 Instruct power controller of new loadings
2 Check power availability		13 Check effect of putting item On/off	6 Instruct cell-room personnel of new loadings
3 Check unit availability		14 Put cell-room on Class 3 level (standby)	7 Instruct user plant of new loadings
4 Check user availability			8 Check consumer plant conditions
11 Check existing loading limitation			9 Balance handling plant

provided in response to this were generally consistent across the different controllers interviewed. When these sequences were then examined, it was apparent that the operations clustered in ways which, on reflection made good sense. Table 3.1 shows these clusters. Reflection upon the clusters, together with further discussion with staff suggested the general categories of *reviewing resources*, *dealing with faults*, *alerting personnel* and *changing the system*.

It was now possible to use these clusters as a basis for a new organisation for the task analysis. This is shown in Figure 3.16 which provides no more information about the task than does Table 3.1. However, it does prompt the need to state the plans and was, therefore, useful in investigating the chlorine controllers' actions further.

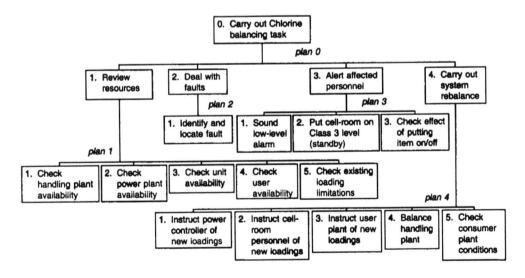

Figure 3.16   An interim revision of the chlorine balancing task analysis.

By stating plans for the task as represented in Figure 3.16, it became clear that nowhere was recorded where the controllers made their initial judgement concerning what sort of response to the basic event was warranted. This step is inferred by the analyst because the controller needs to decide early on whether the disturbance constitutes a 'warned increase', a 'warned decrease' or a 'sudden decrease' in a resource; this decision is fundamental because it guides subsequent plans. It was also apparent that three of the four main operations - 'deal with faults', 'review resources' and 'carry out the rebalance' - were all concerned with affecting the actual changes to the system. With these insights, it now made sense to make a further modification to the analysis, as shown in Figure 3.17. With this revision, plan 0 could now be stated simply. Plan 3 could also be brought into focus.

By focusing on plan 3 it became clear that more was involved in 'establish new resource balance for system' than the three subgoals of plan 3 set out in Figure 3.17. A further revision, presented in Figure 3.18, expands the initial 3 subgoals of Figure 3.17 to six subgoals and their plan. This plan, it can be seen, relies on the fact that the controller has already categorised the event as 'warned increase', 'warned decrease' or 'sudden decrease'.

The complete analysis is shown in Figure 3.19. Again, some of these plans were best represented as flow charts. Plan 3.1, concerned with determining the feasibility of improving productivity, offered some interesting problems. This operation was only ever undertaken when the event was a 'warned increase' in a resource. This was a key decision-making operation. Carrying out this operation well or badly would substantially influence the profitability of the system. Some controllers were more effective in doing this than others. It made no sense to set out these decisions in detail in terms of a flow chart because of the variations possible.

A controller who was offered a warned increase in unit availability would need to determine whether this offer of a resource could be used. To use extra units to make more chlorine, the controller would have to establish whether there was extra power available to enable the additional electrolysis, whether there was sufficient capacity in

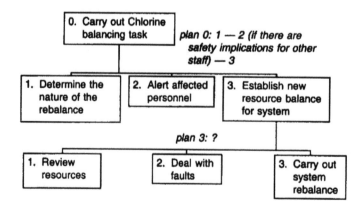

Figure 3.17   Further revision of the top level of the analysis of the chlorine balancing task.

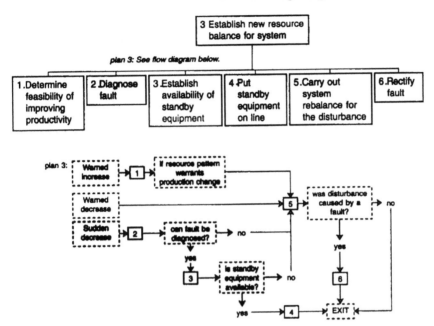

Figure 3.18   Re-examination of 'Establish new resource balance for system. Plan 3 is expanded as a flow diagram.

the handling plant to treat the extra chlorine and whether there were users who could make use of the additional chlorine. If when checking these other resources, anything was missing, then the offer of the extra units could not be accepted. The decision skills that controllers could deploy here were quite subtle. If the controller was offered extra units, then found that there was insufficient power to use them, then nothing should be done. If however, extra power suddenly became available, then the controller would need to confirm treatment plant availability and user availability. However, if an offer of extra power had to be declined because of lack of user availability, the controller would need to check that the power was still available. It must be appreciated that this task, as with most tasks, is part of an ongoing enterprise. Working memory of information states is a significant factor in efficient execution of tasks.

To help explore these issues systematically, a table was prepared to represent the different resource patterns. This is shown in Table 3.2. There are 32 rows, each representing one of the 32 possible warned increases. The 'disturbance' column shows the four different types of resource offer - power, handling plant, production units and users. The 'resources' columns show the different patterns of availability of the other resources. So, for example, row 1 shows a warned increase in power accompanied by availability in cell units, users and handling plant. This means that the extra power can be used to increase production, so this is recorded in the final 'decision' column. Row 2 shows a warned increase in power accompanied by available cells and users, but insufficient handling plant capacity to treat any extra chlorine produced. In this case, then, production cannot be increased and the offer is 'left'.

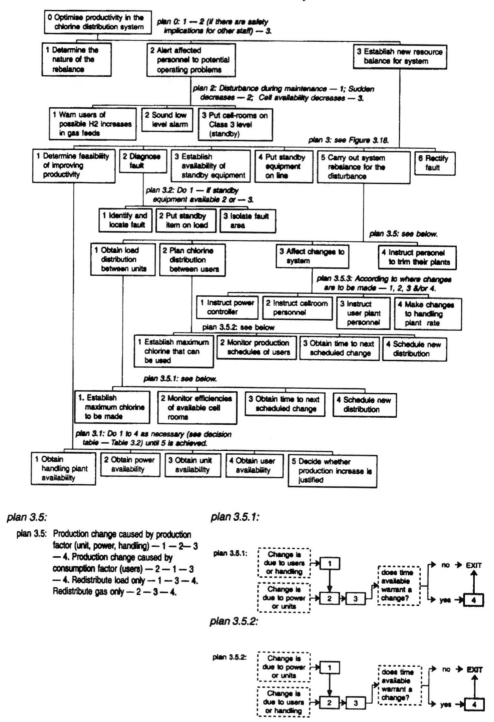

Figure 3.19   The revised task analysis for the chlorine balancing task.

The 32 different cases were explored, both by reference to the logical structure of the system and by reference to controllers and their managers. To accept offers of both power and handling plant capacity always required availability of each of the other resources. Offers of additional cell units were more subtle. Row 17 shows that where all other resources were available, production could be increased. However, if some of the other resources were unavailable it was still worth taking action because the newly offered cell unit might be a more efficient producer of chlorine than those already on load. An interesting exception was row 20. If only additional power was missing, the controller might be able to use the same amount of power more efficiently and, therefore, produce more gas. If there was a warned increase in user availability, the availability

Table 3.2   *Event table for the chlorine balancing task.*

	DISTURBANCE	RESOURCES (withheld until requested)				
		Power	Cell units	Users	Handling	DECISION
1	Warned increase - power	•	•	•	•	Increase production
2	Warned increase - power	•	•	•		Leave
3	Warned increase - power	•			•	Leave
4	Warned increase - power	•		•	•	Leave
5	Warned increase - power	•	•			Leave
6	Warned increase - power	•		•		Leave
7	Warned increase - power	•			•	Leave
8	Warned increase - power	•				Leave
9	Warned increase - handling	•	•	•	•	Increase production
10	Warned increase - handling	•	•		•	Leave
11	Warned increase - handling	•		•	•	Leave
12	Warned increase - handling		•	•	•	Leave
13	Warned increase - handling	•			•	Leave
14	Warned increase - handling		•		•	Leave
15	Warned increase - handling			•	•	Leave
16	Warned increase - handling				•	Leave
17	Warned increase - unit	•	•	•	•	Increase production
18	Warned increase - unit	•	•	•		Redistribute load
19	Warned increase - unit	•	•		•	Redistribute load
20	Warned increase - unit		•	•	•	Increase production
21	Warned increase - unit	•	•			Redistribute load
22	Warned increase - unit		•	•		Redistribute load
23	Warned increase - unit		•		•	Redistribute load
24	Warned increase - unit		•			Redistribute load
25	Warned increase - user	•	•	•	•	Increase production
26	Warned increase - user	•	•	•		Redistribute gas
27	Warned increase - user	•		•	•	Redistribute gas
28	Warned increase - user		•	•	•	Redistribute gas
29	Warned increase - user	•		•		Redistribute gas
30	Warned increase - user		•	•		Redistribute gas
31	Warned increase - user			•	•	Redistribute gas
32	Warned increase - user			•		Redistribute gas

• signifies that resource is available

of all other resources (row 25) would result in increased production, otherwise production could not increase. However, the newly available user might be one whose weekly production targets had been affected from being off-line and so the controller would need to consider redistributing the load between the user plants. Thus, Table 3.2 summarised the different sets of condition that could arise and indicated appropriate courses of action. Generating this table was extremely beneficial because the implications of several of the options had not occurred to management and other staff.

Plan 3.1 could now be stated. The subgoals were stated as 'obtain' the respective resource availability'. The plan states that the controller needs to do 1 to 4 as necessary in order to obtain sufficient information for a decision to be made. This accounts for the fact that on some occasions, the controller need only check one of these to know that a resource cannot be accepted; on other occasions the controller can be more or less efficient in the order in which information is collected given knowledge of what has gone before.

Event tables such as this are extremely useful when stating plans, because they help the analyst avoid being dogmatic about how operators behave. The plan and the table set out conditions with which the operator's strategy must comply, but they do not specify how such tasks must be done.

By persevering in the attempt to represent as much as possible of the task, in terms of a hierarchy of plans, a number of lessons emerged. First, pursuing the disciplines of HTA caused informants to be far more focused on the task detail as planning became explicit. Thus informants were able to rephrase some operations into more suitable forms. Second, by understanding the strategic relationship between different parts of the task, that is, how the different parts of the task contributed to the whole, it was possible to identify other operations that needed to be included and some of the original operations that were redundant. Third, it is clear, as in so many other task analysis projects, that the analyst has to engage in understanding something of the task and its context. It is difficult to conceive that HTA can be completed successfully without the analyst making an effort to understand something of the technology, the underlying working goals and constraints on operators. HTA is not a passive reflection on work activity. It actively engages the analyst in trying to make sense of what is going on.

Finally, the example provides an important demonstration of the power of plans. Together, plans can combine to account for very complex tasks which seemed, at the outset, intractable.

Concluding comment

The purpose of this Chapter has been to set out the importance of plans in HTA. Plans are not simply a logical complement to goals. They contain essential information about how parts of the task need to be coordinated for overall goals to be achieved. Moreover, they combine in useful ways, such that hierarchies of relatively straightforward plans may account for complex operator behaviour.

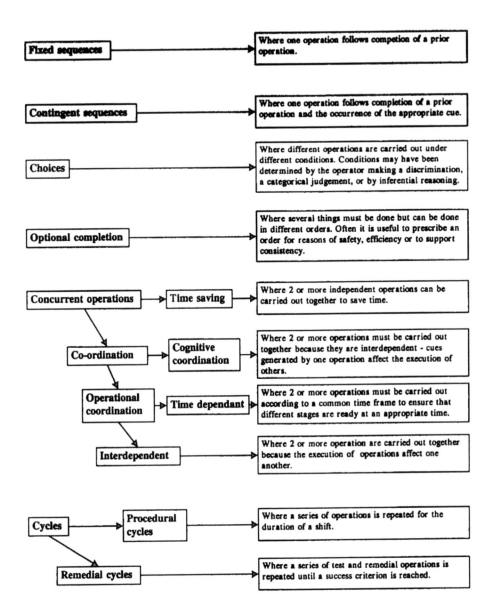

Figure 3.20   Summary of plan elements.

Plans in HTA should always be developed on their merits to represent what the task requires, even though similar plan elements occur in different task analyses. The main plan elements discussed in this chapter are summarised in the diagram in Figure 3.20. These elements do not necessarily account for entire plans, but features within them. However, these plan characteristics can guide the analyst in describing plans in unfamiliar situations. In this way the analyst can account for a wide range of activities in terms of structural elements that appear time and again.

The manner in which plans combine can also be used to tease out complex plans and plans which appear intractable. This was demonstrated by reference to a complex resource balancing task. By examining how experienced operators organised their responses to certain events, it was possible to evolve a hierarchy of relatively simple plans which could deal with a wide range of contingencies. The example of the chlorine balancing task and several of the other examples showed how the need to express plans helps the analyst to identify decision-making.

The next chapter will examine further these issues of decision-making in connection with how practical skills generally need to demonstrate flexibility.

Chapter 4

Flexibility, constraint, cognition and context

Most jobs require staff to be flexible to some extent. This may entail being flexible about how specific actions are carried out or it may require the operator to undertake strategic planning and decision-making. It is also important to acknowledge that performance is also constrained by various factors, such as constraints on how systems are required to work. By encouraging a systematic examination of the task, HTA helps identify requirements for flexibility, the cognitive operations that operators carry out in order to work flexibly and the context in which these cognitive operations are carried out.

Introduction

In some respects HTA appears dogmatic and prescriptive, despite the fact that, as discussed in previous chapters, plans are able to represent tasks in very varied ways. In this chapter we shall consider issues of flexibility, constraints on how tasks are carried out, and how cognition needs to relate to the context in which the task is carried out. HTA helps the analyst recognise when these causes of variation may affect human performance.

Flexibility

All jobs require people to behave *flexibly* to some extent. Doctors and nurses need to vary how they care for and treat individual patients, because all patients are different and their responses to treatments differ. Managers, engineers and supervisors must adapt their activities to circumstances. Events may occur which are not anticipated and with which staff are unfamiliar. Industrial operatives, maintenance staff and clerical staff all encounter situations outside of the expected and, yet, are still required to perform with credit. Indeed, their job may be generally concerned with dealing with the unscheduled or unexpected, rather than routine predictable events, so they must plan and make decisions. Office workers are more valued if they can remember procedures that worked on previous occasions. If a salesperson in a bookshop is able to make sense of a vague enquiry, then a sale will have been made. Moreover, the customer is happy and impressed and will come again. Even in more straightforward manual tasks people have to adapt. They may need to adapt as they get tired, as the characteristics of the materials they are dealing with vary and as they tools they are using wear down or change their operating characteristics. Thus, people may need to press harder, inspect their work more frequently, be more patient, take a rest, reject unworkable materials, and so on. This need for flexibility may be assumed to be at odds with the rigorous and prescriptive nature of task descriptions using HTA. Appreciating the need for flexibility is an important part of task analysis.

Constraints on how tasks are carried out

Just as all jobs entail some degree of flexibility, so all jobs are *constrained* to a greater or lesser extent. They are constrained with regard to the system's goals that the operator is expected to attain. People are employed so that they work towards the organisation's goals. This means that the person is expected to adapt behaviour to ensure that goals are attained. Many jobs are also constrained by the methods that must be adopted. In many jobs, some parts of the task must be done before others, to ensure that hazards are avoided, satisfactory manufacturing is assured and colleagues know what one another is doing at a particular time. Some jobs are rigidly prescribed. This is often the case where substantial risk is involved. For example, operations carried out in the pharmaceutical industry must often comply with given standards as must many health practices. This means that discretion is limited in many areas.

Individual differences and variations

Even where different people appear to do the same thing, they may be doing it in a different way, using different strategies and different psychological resources. On confronting a particular set of circumstances, one person might take stock and think through what needs to be done. Another person may adapt a response to a similar situation. Another person might make an inspired guess. Different people have different experiences and have learned to organise their knowledge about the world in different ways. Yet each person appears to be achieving the same objectives.

Variations also occur within the individual. If a person realises that something they are required to do is the same as something they have done recently, they will be far more fluent in their response the second time around, provided the first encounter was successful. Equally, had the first encounter proved problematic, they will be more cautious on the second occasion. Furthermore, if something is perceived as urgent, then a risk might be taken. If an operator is getting tired and is concerned about making mistakes, then greater caution might be exercised. People change with circumstances and people differ from one another in view of their personal histories and factors such as their individual tendencies to take risks or the extent to which uncertainty causes them to become stressed.

Cognitive task analysis

This flexibility of human behaviour has prompted some writers to champion methods of *cognitive task analysis,* which aim to explore the ways in which people deploy cognitive processes when they carry out tasks. Cognitive task analysis methods include methods for examining behaviour and for representing the way task knowledge is organised and the processes through which operators deploy knowledge and skill to demonstrate their expertise[21]. Indeed, cognitive task analysis methods may reveal facets of human behaviour which HTA does not entertain. There is a danger that such methods do not fully account for the task context in which applied cognition is situated with the consequence that they investigate cognition for its own sake without reference to where operators need to concentrate their effort.

A common view is that some tasks warrant a cognitive approach while other tasks should utilise a non-cognitive approach. Part of the difficulty in distinguishing between cognitive task analysis methods and other forms of task analysis is that the analyst is required to make a prior judgement on whether a task is 'cognitive' or whether it is not. This view is flawed because *all* tasks rely on processes of cognition to ensure they are carried out effectively. They require information to be monitored, an appropriate response to be selected and the consequences of actions to be evaluated to ensure that actions are adapted and controlled appropriately.

While there are neither wholly cognitive nor wholly non-cognitive tasks, we might refer to those tasks where the operator must interpret lots of information in order to monitor system health, diagnose system states, or plan actions, as 'cognitively loaded tasks'. Even for such tasks, an evenhanded strategy should be adopted in which the *need* to examine cognition emerges as part of a general task analysis process. Otherwise, the analyst may be engaged in substantial futile work. For example, a task involving human computer interaction may be extremely difficult, but the difficulties may be resolved by rescheduling work demands, or providing a more supportive learning environment, without undertaking an extensive and expensive examination of the behaviour of the operator trying to work in a poorly designed task. Driving instructors and music teachers are both concerned with teaching skills that require careful judgement but their success does not depend upon having a full appreciation of the cognition involved. They make progress by setting their students goals and providing opportunities for guided practice. To do this, they establish what the learner should achieve and then devise strategies for helping the learner master the required skills.

It is also important to remember that task analysis is undertaken as part of an *applied* intervention. This emphasises two important considerations. First, cognition is affected by the *context* in which it is deployed. Unless context is understood, the ensuing analysis may be ill-informed because the performance may be affected most critically when certain factors come into play and may not be affected when the context is benign. Thus, dealing with a diagnostic problem may be observed to be straightforward when other aspects of the workplace are well under control; problems may only arise when staff are stressed by other tasks. The second practical consideration is criticality. A task element may be difficult, but if the consequences of error are trivial, then time and other resources to analyse that element are not justified.

We should also remember that, strategies for examining and representing cognition are not themselves particularly valid in the sense of leading to reliable and meaningful results which apply to all people. Many of the methods used to collect data in cognitive task analysis, such as observing performance, analysing verbal protocols, and examining this data from a cognitive perspective, are speculative and do not necessarily provide an account of cognition with proven reliability. But they are useful as methods for engaging the analyst with the task. Indeed, they may be best used within a general approach to task analysis, such as the HTA framework. Their role within such a framework would be to help with the process of generating design hypotheses.

HTA can provide a number of practical benefits that can help the analyst examine tasks where flexibility is evident or required. The following issues will be described briefly and then discussed more fully.

- Inferring cognitive operations
- Focus and bias
- Modelling and evaluating strategies
- Situating cognition
- Inferring cognitive skills and resources
- Identifying strategies off line

Inferring cognitive operations

Cognitive operations are those operations that are driven substantially by cognition in order to make complex decisions, rather than simply to guide actions. Cognitive operations include problem solving and system monitoring. Much of what we understand by 'cognition' in task performance is covert and has to be inferred by the analyst. For example, it is impossible to judge that someone is making a decision simply by observing their behaviour as they may appear to be doing nothing. Decision-making is often inferred from considering how a person's performance varies subsequently and by considering the demands made on the operator by the system under control. The analyst must, then, be able to infer when a flexible response, dependent upon a cognitive skill, is required. The evidence for such behaviours often emerges when *plans* are written.

Focus and Bias

At the start of any project, the analyst or the client may have preconceived ideas about the problem. There is a danger that an analyst could actively seek complexity and promote the idea that people need to be flexible when, in fact, this is not required. Also, in order to make the task appear more straightforward, some analysts or their clients might wish to overlook the fact that flexibility and decision-making is required. It is important that task analysis methods help the analyst to focus on real issues and deal with those issues where attention is justified.

Modelling and evaluating strategies

There are occasions when several alternate strategies or methods may be used to attain a goal. In some cases freedom to devise a method may be given to the operator. In other cases there are constraints on which methods should be adopted. Some strategies are more memorable than others. Task analysis can be used to identify alternative strategies as a basis for this sort of evaluation.

Situating cognition

Someone may carry out similar cognitive operations on different occasions in different contexts which give rise to different conditions. For example, a particular operation could be carried out while something else is happening or it might be required unexpectedly or it could be fully anticipated such that preparations can be made. It is important that the analyst understands the conditions under which such cognitive operations are undertaken.

Identifying strategies off line

In many cases, procedures and decision-making are identified prior to the task being carried out under real operational circumstances. As the task is analysed, those aspects where the operator is required to perform flexibly are identified, as are those situations where the operator must respond according to an approved procedure. These decisions are made in safety-critical situations where managers and engineers have to prepare procedures for operational staff to follow.

Generating design solutions

Even where an operator is required to demonstrate flexibility, it does not follow that the underlying cognition needs to be understood in any formal sense in order to devise solutions to improve performance. Often, an operator can rehearse flexible strategies on-the-job or in a simulation and, thereby, demonstrate flexibility. This is what happens in the case of learner drivers and music students.

Inferring cognitive operations

By 'cognitive operations' is meant those mental operations when operators use information gleaned from an interface in conjunction with their own knowledge, in order to make decisions and choices concerning how to proceed towards a working goal. Whereas actions the operator takes and, to some extent, the information that the operator collects, can be recorded reliably by careful observation, *cognitive operations* must be *inferred* by considering the wider pattern of actions undertaken and relating this to the operator's intentions.

The presence of cognitive operations is often first suspected during interviews with operators or through the analysis of verbal reports in which the operator states that a cognitive operation - such as making a decision, solving a problem or monitoring the state of the system - is carried out. Such claims are important leads, but they need some validation because people may be elaborating parts of their job. Validation can be achieved by considering the consequences of decisions. Thus, if different sets of actions are consistently carried out in different circumstances, then a decision process can be inferred to account for the operator's choice. This is the case even where decision-making was not apparent through observation or when talking to the operator. If a checkout operator in a supermarket consistently asks mothers with children or people with some kind of infirmity whether they would like help with their packing and yet does not ask the same thing of other people, it can be inferred that a *discrimination* has been made even though this is not obvious to an observer.

Similarly, if following a period of apparent inactivity an operator is seen to respond swiftly and appropriately to an unforeseen set of circumstances, then the operator can be assumed to have been *monitoring* a signal. In many task analysis projects it becomes clear that monitoring by the operator is required as a key method to detect deviation from system goals, whether or not this is apparent from observing operators at work.

The process of inferring the presence of a cognitive task element within the context of a task analysis is assisted by identifying what the task is supposed to achieve. A monitoring or decision-making operation may vary considerably in accordance with demands placed on the person carrying out the task. This may be illustrated by reference to the case of a railway systems engineer, employed to monitor the health of the railway system in terms of trains running to timetable and whether assets (trains, track, escalators, etc.) are functioning correctly. If a problem arises the engineer needs to decide on a course of follow-up action.

This task is undertaken in a central control room, with maintenance teams deployed to undertake the work at the affected site. The demands on the monitoring and decision-making undertaken by the systems engineer will vary in accordance with the organisation of work. In one form of work organisation versatile technicians may deal with all the technologies of the railway - mechanical, electrical, electronic. The systems engineer would undertake an initial diagnosis, then brief a maintenance team. The initial diagnosis cannot be complete, because the central control room only enables limited access to information; inspection of local conditions is also necessary and this must be done by the maintenance team when they reach the site. Therefore, the engineer's task is one of reduction of uncertainty to facilitate the search that the team will need to carry out when it reaches the site.

In a second form of work organisation maintenance teams are not multi-skilled but specialize in particular technologies. Here, the engineer must deploy the team most likely to deal appropriately with the problem. Deploying any team is costly. Should it transpire that the initial judgement was wrong, then these costs will have been wasted and the next hypothesis selected. Moreover, the judgement of whom to deploy will be a function of availability, costs, likely search time, likeliness of solutions and the severity of the incident. These judgements will be far more demanding than the simpler hypothesis reduction needed in the first form of work organisation where teams were multi-skilled.

A further variant is that contractors are used to deal with all maintenance; the systems engineer merely monitors the contractors' response to events. This entails noting incidents, then awaiting telephone calls from the contractors' management to say that their teams are responding. The engineer's job would then simply be one of quality monitoring to ensure that the terms of the contractors' contract are being observed. Each of these variants entails monitoring and decision-making, but associated cognitive skills will be trivial or complex in accordance with the decisions and actions that follow. Any task must be first understood in order to anticipate the demands placed on cognition.

Just as consideration of actions helps the analyst to identify the cognitive operations involved, so identification of cognitive operations points to actions to be included in the task analysis. An operator may report that, after '*checking* the melting point of a substance', a standard fixed procedure is followed. Observing several instances of the task may appear to confirm that this is a fixed procedure plan. However, *checking* or any other activity concerned with collecting information implies at least two outcomes - that the system status is *satisfactory* or it is *unsatisfactory*. There is always a danger that observing an invariant routine merely reflects that the system is reliable and that unacceptable outcomes are rarely encountered. If, on the other hand, remedial action is never necessary, then the checking action can be ignored. The only occasions when checking a parameter is a legitimate part of a fixed procedure is when the outcome is recorded for use by somebody else, for example, quality control. Task analysis should always establish the interaction between cognition and action before making assumptions about the nature of cognition or, indeed, the necessity for action. This is done by writing satisfactory plans.

Focus and Bias

Since real performance problems are complex there are often several factors to which the problem can be attributed. This can bias the investigation. There is a danger that the analyst or client focuses upon an area of particular interest to themselves. Cognitive operations may be particularly interesting to an analyst who may be tempted to give them undue attention. The client may prefer to focus on an unfamiliar aspect on the assumption that this must contain the root cause of any problems being experienced. Sometimes managers favour a particular type of solution to a problem because it is less intrusive than other types of solution. For example, a manager might require that *personal selection* is dealt because it avoids making costly and intrusive engineering solutions.

An example of management focusing on the unfamiliar was concerned with a computer-based accountancy task in a large service organisation in which operators were concerned with using computer terminals to set up new customers and deal with data entry and retrieval. The management had little experience of their staff using computers in such numbers and decided that these aspects of the task, were the most likely candidates for attention. Indeed, the computer screen design was uninspiring and often inconsistent, so the analyst was tempted to dive straight in and focus on these aspects. However, a broader analysis placed these human computer interaction (HCI) issues into perspective. Despite the presence of computers, the broader task analysis showed that the real problem resided in the manner in which clerks were required to investigate problems with customers over the telephone, rarely involving use of the computer. Despite their poor design, it was clear that operators could learn the computer tasks effectively and that errors were easily recoverable.

In another example, management of a process plant assumed that a new computer-based plant display would present their experienced operating staff with no problems as they came to adapt to new working arrangements. Task analysis was seen as a method of analysing and designing familiarisation training. In carrying out the task analysis it was revealed that mastering the new version of the task was far from being a simple familiarisation exercise. It was shown that the configuration of information on the screens was inconsistent with what the operators were required to do. Thus detailed analysis of the cognitive skills involved in human-computer interaction was justified in order to facilitate screen redesign.

Each of the examples given was a practical problem facing the management concerned. Each warranted some form of task analysis. In the first case, expectations created the risk of unnecessary attention to cognitive aspects of the task. In the second case, the converse was apparent. Both cases benefited from taking an evenhanded approach in which the cognitive operations were understood in the wider context.

Modelling and evaluating strategies

Most tasks can legitimately be done in several ways. Tasks which entail diagnosis in complex situations are often so rich in information that they can be accomplished with very different strategies which vary subtly or substantially. Some people might be content to pursue a strategy of pattern recognition where performance is based on comparing current circumstances with known patterns of symptoms or events. Other people might prefer to reason about symptoms. Others are able to recognise when a rapid pattern matching strategy is suitable and when a more reflective approach is required[22]. In some cases, it is necessary to impose constraints on how tasks are carried out. This can be done by using plans to model and evaluate different strategies.

Figure 4.1 represents part of a dispatch preparation task in a computer controlled warehouse. Items in stock are contained in product hoppers. They are used to make up specific orders for retail outlets across the country. Details of each order are input to a computer which then controls gates at each hopper, opening the gate to drop the required number of packages onto the conveyor belt. The operator is employed to identify the end of one order and the start of the next. The operator is supplied with a batch of dockets containing dispatch details of each order. The operator has a visual display terminal which shows the code letter of the first and last 3 items of each order. Each package has its code printed on it. When the gap is identified, the operator places a plastic tray on the conveyor belt and drops the appropriate dispatch docket into it. The operator must keep up with the pace of the conveyor belt which is moving quite fast. Carrying out this task effectively is important because it provides the transport supervisor with the necessary information necessary to load vehicles for dispatch.

At first glance, this is a simple task. The operator looks for a gap, then read the codes on the packages either side and confirms this with the information on the screen. Problems arise because packages may fall onto the sides on which their codes are printed, so they cannot be easily identified. Also, gaps may arise because items are out of stock

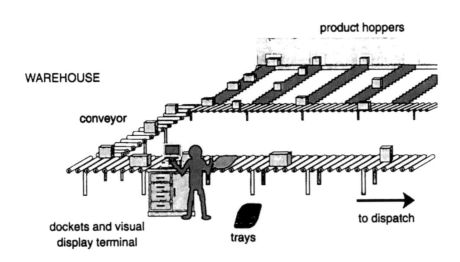

Figure 4.1   Representation of a conveyor belt operation in an automated warehouse.

and have not dropped onto the belt. Given that the belt is moving at some speed, these factors can promote difficulties if the operator were to identify a gap and then find the product codes difficult to locate. Gaps between orders may be confused with missing items. The operator can adopt different strategies for doing this.

In principle, the operators could be left to deal with the task as they wish. However, by using HTA we set out two alternative strategies which were likely to give rise to different likelihood of error. Figure 4.2 shows one variant. For each gap, the operator first establishes the docket required for the next batch. Normally, the next docket is taken from the top of the pile, but if one is missing, the operator must insert a 'location missing docket' as in subgoal 2.4. - the task of matching the order to destination is then left to the dispatch department. Then, goal 3, 'locate correct gap between orders', is carried out. In this version of the task, the operator first reads the product codes surrounding the target gap on the VDU and commits these and their position to memory. Then the operator looks along the line to identify the next gap for comparison. This is difficult because the codes might be obscured on some items. If a match cannot be made, the gap is rejected and the next gap sought for making the comparison.

Figure 4.3 represents an alternative strategy for accomplishing the same outcome. In this strategy, the operator first identifies a potential gap, then scans packages either side of this gap to locate identifying codes. Then the operator looks for the equivalent positions on the VDU screen to confirm the match. The strategy in Figure 4.2 is far more dependent on remembering the six items that surround the target gap as displayed on the VDU. In Figure 4.3, the operator scans to locate a gap then commits to memory

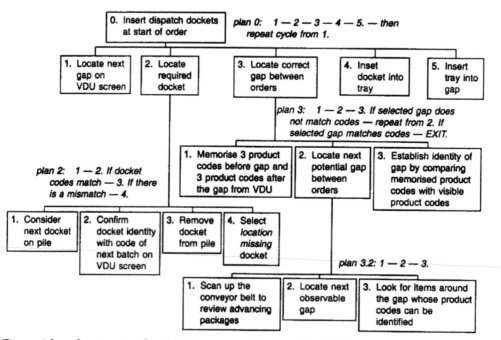

Figure 4.2   One strategy for identifying spaces for inserting dockets.

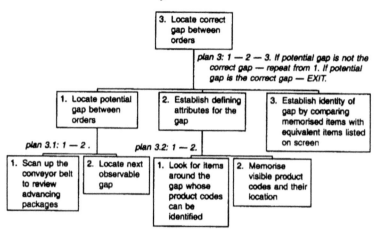

Figure 4.3   A second strategy for locating the correct gap between orders.

only those items whose codes are accessible. It is a reasonable assumption that the strategy in Figure 4.3 is likely to be freer from error than that represented in Figure 4.2. A more reliable judgement would be made if the analyst were to carry out some simulation trials to see which strategy really is most robust.

This is generally a badly designed task. The best solution would be to redesign the software, but this is something that management did not wish to entertain. Consequently, the alternative was to identify the best strategy for people to follow as a basis for operator training.

Situating cognition

A problem facing all human factors analysts is understanding the effects of the wider context in which a task resides. Simply identifying a type of behaviour is not a sufficient basis for recommending how it should be treated. There are several ways in which the task context can affect how behaviour can be influenced, including:

- goal context
- frequency, predictability and coincidence
- priming and sharing information
- decision outcome.

The hierarchical structure of HTA helps us to understand something of these influences. This will now be discussed.

Goal context

Goal context refers to the wider goals within which a task element of interest resides. This issue is illustrated by the HTA of the task of detecting flaws in welds, using ultrasonic probing methods, in the manufacture of pressure vessels for the nuclear industry[23]. Where welding is used to bond metal together, there is a risk that a weld may be flawed and not withstand the forces that will act upon it. Thus, it is incumbent upon people operating risky processes, for example railways and aircraft, to demonstrate that their welding is free from such flaws. A method for doing this is to pass ultrasonic signals through the objects containing welds, then measure the reflection of the signal. Flaws may be detected by judging how erratic the reflected signal is. One context in which this is done is in the production of pressure vessels, used to contain materials at high pressure. Such vessels usually contain welds in order to fit pieces which allow the vessel to fit onto other items of plant. This is of particular concern in nuclear processing, where unscheduled emissions of radioactive materials under pressure can be catastrophic.

Figure 4.4 shows extracts from the HTA of the task of detecting flaws in welds, using ultrasonic probing methods, in the manufacture of pressure vessels for the nuclear industry. The task entails carefully moving an ultrasonic probe (something like a mouse

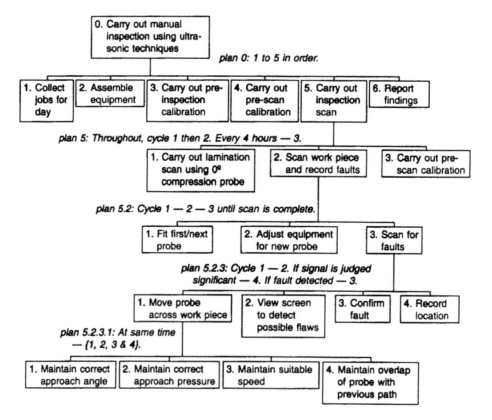

Figure 4.4   Extracts from the HTA of ultra-sonic testing of welds in the nuclear industry.

on a personal computer - see Figure 4.5(a)) across the surface of the vessel being tested. The area being inspected is covered by a jelly couplant to ensure best transmission of the signal. The operator's job is to maintain a consistent pressure, ensure all parts of the area are inspected and, at the same time, the operator must monitor a cathode ray tube for evidence of a flaw - see Figure 4.5 (b). At the best of times, the cathode ray trace is very erratic or 'noisy'. The operator must distinguish between occasions when the display indicates a flaw from those occasions when there is presumed to be no flaw. This is not easy, especially because it depends, in part, on the consistency of maintaining a constant pressure on the probe.

The HTA shows how the different operations in this task need to be carried out in conjunction with one another. Maintaining consistent pressure, monitoring the trajectory and monitoring the screen to detect a flaw are all important skills. Unfortunately, they cannot be trained separately because they need to be coordinated - the skill is doing them effectively together. One aspect is that if the operator detects a possible flaw, but is aware of a momentary loss of concentration on pressure, or knows that the geometry of the vessel has interrupted the normal flow, then a possible explanation was that the probe was not controlled properly, thereby causing the disturbance to the display. So the operator would re-test that part to check the result.

This task of identifying signals in a display which is normally noisy is a 'signal-detection' problem[24]. Here, 'noise' refers generally to random background stimulation. The difficulty for the operator in these situations is deciding whether an apparent signal is a proper signal or whether it is simply additional noise. Equally, if a real signal

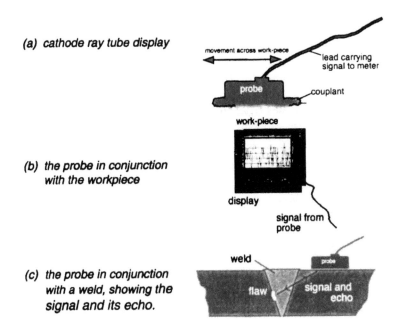

(a) cathode ray tube display

(b) the probe in conjunction with the workpiece

(c) the probe in conjunction with a weld, showing the signal and its echo.

Figure 4.5   Equipment used in the ultra-sonic testing task.

occurs when the noise is far less extreme, the operator might judge this to be noise on its own and, therefore, fail to detect the signal. In tasks such as these, we assume that the operator adopts a threshold below which the signal is judged not to have occurred and above which the signal is judged to have occurred. This means that there will be situations when a real signal is missed and situations when a signal is recorded which has not occurred. These two different sorts of error have different consequences. In this ultrasonic testing task, falsely identifying an absent flaw will result in inconvenience for other staff who would need to check the work and it could mean that an expensive vessel is rejected when it was perfectly good. On the other hand failing to detect a flaw can result in a faulty vessel being commissioned with the a possible consequence of rupture and radioactive emission. It is in the interest of public safety, therefore, that operators adopt a conservative strategy, even though they will report more welds than there actually are because the consequence of the alternative cannot be countenanced.

Where people are required to make judgements of this kind their judgement threshold will be influenced by a number of factors, some of which will not be obvious. One set of influences could be the employment context. From a commercial perspective there is a concern to complete an inspection programme as quickly as possible to avoid delays to construction. A management wishing to encourage this could introduce a bonus scheme, with bonuses contingent upon the performance of teams of inspectors completing their inspection programmes on time. In this event operators may feel a social pressure to avoid jeopardising the team's bonus and, therefore, be less inclined to reject welds that might seem faulty in other circumstances. Whether or not inspectors were conscientious or cynical, their individual performance may well be affected unconsciously either by the desire to secure a bonus, or to avoid the social stigma of jeopardising the team's bonus. Clearly, a bonus system such as this is not consistent with maintaining a conservative inspection policy.

The case described is typical of the problems facing management in high risk industries or the health services. It is not meant to suggest a cavalier attitude on the part of management. Indeed, elsewhere, the steps suggested would be construed as good management practice and would be favoured by the workers themselves. However, there are hidden consequences of such management strategies which can be brought out by task analysis. These substantially affect the working context in ways that will influence performance.

Much of the insight into this problem derived from considering the nature of the work encountered in terms of wider human factors theory. It did not derive explicitly from the task analysis, although the task element of concern was identified through the task analysis. In addition, the wider task context was traced back through the task hierarchy such that the influence on specific parts of the task or factors concerning the overall job could be understood.

Frequency, predictability and coincidence

Further examples of the importance of understanding individual task elements in their wider context relate to issues of task demand and timing. These relate to the incidence of events that trigger action. It is helpful to think in terms of three main factors:

- frequency of event

- predictability of event

- coincidence of events

The *frequency* of an event will affect both the opportunities that operators have to rehearse activity and the extent to which investment in a solution is justified. If an event is *predictable* staff can make preparations to cope with it. Management might strengthen work teams at the time when the event is anticipated, and the individual operator might be able to organise work more effectively and gather necessary tools and information to deal with the imminent event. If events *coincide,* this can add to workload. These factors can combine. For example, if infrequent events occur together in an unpredictable way, the operator may be unable to cope. It is the concern of the analyst to understand the nature of the operations that will be used to deal with these various sets of circumstances. The key point is that the operations must be understood in the context of these combined effects.

Frequency and predictability

A good example to illustrate this point is that of diagnosis. Diagnosis is critical in the control of many systems. When a system goes off specification, it can no longer fulfil the purpose for which it was designed. The operator supervising the system is generally required to monitor the system's health in order to identify threatening events and select contingencies to deal with them.

Where plans contain contingencies - 'if' statements - then issues of frequency and predictability come to the fore. The plan in HTA does not state the frequency and predictability of events, but it does state the events concerned. For example, it will not state how often a particular subsystem fails but it will state that if this subsystem fails, then a given response is required. Where such events are identified, the analyst must then establish their frequency and predictability.

Figure 4.6 shows the HTA of a supervisory control task concerned with maintaining satisfactory service on an automated railway line. Automation in this context means that signalling is controlled by a computer in order to meet the requirements of the timetable and maintain a proper distance between trains. During normal running, the operator's job is to maintain best service. This means that when everything is working properly, the operator will simply monitor that key parameters of the railway remain in tolerance (subgoal 2.1.1) - these include compliance with timetable, and satisfactory headways between trains. The operator also monitors the railway assets (subgoal 2.2) - to ensure that systems are working properly, for example, that signals are working.

The plans in this task analysis contain a number of contingencies. In plan 2.1 problems may be identified for attention because monitored parameters are out of tolerance. Many of these problems will be infrequent and unpredicted. Depending on urgency, there may be a need to train people to deal with these problems unaided or a written procedure might be provided. For problems that have not been anticipated, the operator must be capable of effective reasoning, otherwise the system would have to be

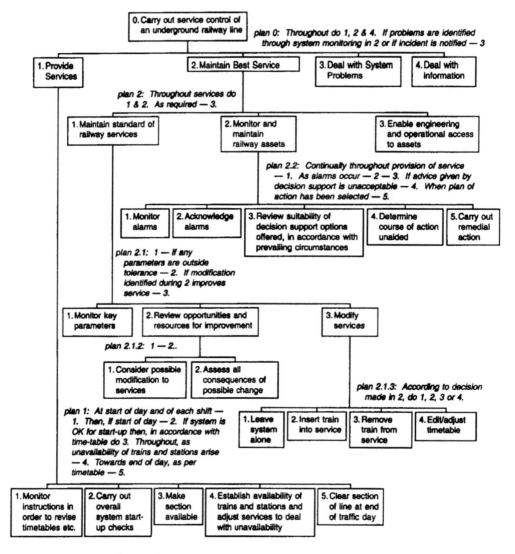

Figure 4.6   HTA of an underground railway control task.

shut down. In contrast, in plan 1 contingencies refer to time of day. This is entirely expected so operators can be given explicit instructions concerning how they should vary their performance. It is also clear from this analysis that in response to some events, the subgoal invoked is an action which should be carried out without variation, while in other cases, the subgoal requires a cognitive operation, such as diagnosing a problem or selecting a suitable course of action.

Coincidence

Where events are unpredictable, several might occur together, causing extra workload. For example, in Figure 4.6, 'Make section available (subgoal 1.3)' could be required at the same time as 'Establish availability of trains (subgoal 1.4)' — this possibility is implied by the plan. Further, since plan 0 requires subgoals 1, 2 and 4 to be carried out throughout, further coincidences might arise. So, while the operator is dealing with the problems associated with bringing trains into service, there may already be problems developing concerning the trains that are already in service. These multiple problems are, of course, issues of system reliability. A highly reliable system should limit these coincidences. A poorly maintained system is likely to increase the operator's workload substantially. These multiple problems bring with them problems of workload and stress and point to job-design solutions to increase the size of operating teams, although this may not always be appropriate.

Priming and sharing information

If the operator has just carried out an activity that always heralds the new task demand, then performance will be *primed*. Some priming is *procedural* in the sense that it prompts the operator what to do next. Indeed, it is common when asking task experts the details about what they do to discover that they are unable to give a clear account. Yet when placed in the workplace, they perform the correct actions because they are prompted by the context.

Some priming is *informational*. Attainment of a previous goal may provide information to the operator, which will be of use in later activities. An operator who is expected to diagnose a problem will perform differently if he or she was previous engaged with the system. Thus working arrangements where one category of staff monitor system performance, then refer the problem to another person for diagnosis risk the diagnostician being ill-informed.

The analysis of the job of a ward nurse demonstrates the role of informational priming and shows how this could have consequences for staffing practices. A staffing issue that concerns the medical profession relates to allocating some domestic duties to non-medical auxiliaries. Whilst considering this task, it was recognised that the nurse needs continually to review the patient's progress and that the main way in which this could be done unobtrusively is while other routine domestic tasks were being undertaken. The case was considered of an elderly patient who was failing to respond to treatment · because of anxiety about who was feeding her cat. Such concerns typically emerge through normal conversation. A nurse who is party to that conversation who is also conversant with the patient's care plan may recognise that this anxiety could be the cause of the poor response to treatment. An auxiliary staff member would be sympathetic, but not have the other information or expertise to make the link.

Figure 4.7 represents another medical context, that of neonatal intensive care. Doctors and nurses must work in close harmony to care for and treat babies whose health is at risk. Of these 10 subgoals, some are done by nurses and some by doctors.

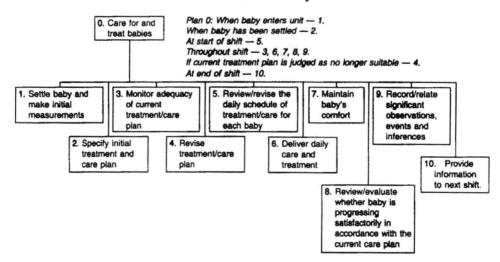

Figure 4.7   The general functions of a neonatal intensive care team.

Their joint effectiveness depends upon how well they collaborate. Of the doctors' tasks those involving general clinical judgements are done by experienced senior staff.

Carrying out detailed treatment procedures are done by senior house officers. Neonatal intensive care units vary in how they organise and allocate duties. However, analysis of these tasks shows how they are interdependent with regard to sharing information. In 'monitor adequacy of current treatment/care plan' (operation 3), for example, staff need to know what has gone on in the past in order to make sense of the present. One nurse reported how she monitored that a baby's oxygen levels which seemed satisfactory until she remembered that the baby had been put on an increased oxygen dosage 8 hours previously. While the current blood-oxygen value was satisfactory, had things been left, the baby would soon be on too much oxygen.

Decision outcome

Just as what has gone before affects how a person carries out an operation, so too can what follows. The issue of *decision outcome* has already been discussed in relation to the railway systems engineer's task. A diagnostic problem where the operator must distinguish between which of several specialist maintenance technicians need to be alerted is very different to one where the operator is required to determine the course of remedial action in terms of which pieces of equipment need replacing, and is very different to a binary choice of keeping a system operating or shutting down. Tasks such as these are best understood within the context of a wider task.

Identifying strategies off line

A debate in task analysis that has been alluded to throughout this chapter concerns whether task analysis methods, which are in some respects prescriptive, are appropriate when considering how people can be made flexible to carry out their jobs. HTA is, in many respects, prescriptive. However, it has been shown how it can be used in ways that inform a flexible approach. A final point needs emphasising. In many real complex tasks operator discretion and flexibility is not sought in operator performance.

In safety critical systems things often need to be done in an approved way. In complex automated systems, solving problems may be too difficult to do reliably under operational time pressures. The normal practice in such industries is for engineers to work out different contingencies, examine them carefully, obtain approval for dealing with contingencies in this manner, then take steps to ensure that operators comply with procedures by developing appropriate training, job-aids and system interlocks, for example. In some cases it is important to do things in this way to obtain a licence to operate. Thus, operators will carry out very complex tasks but will not have generated these safe and effective strategies for themselves. Engineering and other management support staff combine operating experience and system knowledge with time to reflect and away from operational stresses. Much of the cognitive skill expended in carrying out such complex tasks is done off-line. This does not discount the possibility of identifying areas where operators are given discretion to be flexible. But now the conditions for helping them acquire necessary expertise can be addressed. Thus, HTA can be used as an off-line tool to enable the prescriptive and flexible elements of the task distinguished and appropriate methods devised to support each.

Concluding remarks

There are aspects of all jobs in which the operator must demonstrate flexibility. This includes organising commonplace actions to suit current conditions and it includes more cognitive operations such as planning diagnosis and monitoring. There are also many parts of tasks where the operator must act in a prescribed way. Such occasions are frequently encountered in safety-critical tasks where deviation from carefully thought through procedures is wholly unacceptable and risks accidents and fines. In other situations managers and engineers specify strategies to be followed to ensure that product . or methods comply with a required standard. An important part of task analysis is to identify which aspects of a performance can or should be flexible and which aspects must be fixed. For this reason alone it is necessary for a task analysis method to allow the analyst to make such judgements in an even handed way. The HTA framework provides a means for the analyst and client to focus on critical aspects of the task and to make judgements about the status of each part encountered.

When flexible performance is required, it implies some form of planning or decision-making on the part of the operator. Such cognitive operations are usually covert. Task analysis must be used to identify and characterise the nature of these operations.

This does not mean switching analytical strategy immediately to a *cognitive* approach because there is much to be done before considering performance at this level. First, the nature of the cognitive operation can be clarified in terms of its role with respect to the wider task. Thus, *monitoring* operations are identified when it is judged that the operator has to maintain a watch for system deviations, *diagnosis* is necessary when the operator must determine which of several courses of action to follow and planning is necessary when it is judged that the operator has to change the state of the system. Second, the *context* in which the identified cognitive operation is situated can be understood. This indicates the main goals to which the cognitive operation is contributing. It shows where the operator is subjected to additional workload through having to deal with several things at once. It shows the effects of the things that were done previously and what follows later. These contextual factors can be identified be reference to the plans in HTA and the relationship between plans in the task hierarchy. If the analyst then needs to adopt a cognitive task analysis approach, this can now be done, whereas, adopting a cognitive task analysis approach before understanding the context of the operation can be quite misleading.

This and previous chapters have dealt with the rationale and the methods of HTA. The next chapter will deal with the practical issue of representing the completed analysis in order to communicate to other people and to maintain a record of the work done.

Chapter 5

Representing and recording HTA

Maintaining good records is essential both to manage the process of a task analysis and to justify, at a later date, the decisions made. Effective recording methods must account for the complexity of the task in a number of ways and still be relatively easy to follow.

This chapter will discuss why record keeping during and after analysis is important. It will set out methods for presenting diagrams, different tabular formats, and discuss the respective merits of each.

Much of the content of this chapter is practical and may be used simply for later reference.

Introduction

So far in the book, HTA has been represented in the form of hierarchical diagrams. This is probably the clearest way to communicate analyses since it shows how subordinate goals and their plans are organised to represent the goal they are redescribing. Of concern, however, is the tendency for people to assume that HTA is only ever represented through diagrams - that if a task description is represented as a hierarchy, then it is HTA; if the description is in the form of a table, then it is not. This view is quite misleading. HTA, as described in Chapter 2, is a systematic process of reviewing goals in terms of suboperations and plans, and continuing this process until there is no benefit in going further. Thus, representation of HTA must include reference to why further redescription within the analysis was stopped at any point and any assumptions about the context of the work. To provide a satisfactory record of the work, it is generally necessary to use tables as well as diagrams to represent the analysis.

This chapter will describe different methods of representing HTA to fulfil different purposes. The methods discussed have been developed in practical contexts and shown to work. This does not exclude other methods being developed which could equally fulfil the requirements of the task analysis project.

Reasons for representing and recording

Careful task representation and recording is necessary in communicating results to others. As a task analysis progresses, the analyst works with one or more task experts to obtain information and represent it in an appropriate way. There will be occasions when input from other people is sought either to confirm results so far obtained or to fill in detail that has been missed out. It is necessary, therefore, to represent any progress to date so that the task expert or informant may provide further information consistent with what has already been collected.

When a first draft of the analysis has been completed a good record is necessary so that the work can be reviewed and validated. The overall content of the analysis must be checked for accuracy, completeness, and to determine whether it satisfies the requirements of the client. To accomplish this requires that the analysis is critically read by the client or the client's agent. It is also important to ensure that the work can be used by others, for example, people designing training or interfaces or safety analysts.

A task analysis should also provide a record to be kept for future reference within an organisation. In many instances, task analysis needs to be reviewed from time to time. New equipment, staffing changes and new legislation all have consequences for how a task is carried out. A previous task analysis can be reviewed and modified accordingly. If there are incidents which require investigation, having a good task analysis on record will enable a swifter response or provide a justification for a decision that subsequently proved problematic. Equally, decisions taken about the task during the earlier task analysis may need to be reviewed.

Where tasks are complex, task analyses can become large and work may need to continue over days, weeks or even months. Where work on a task analysis is interrupted for even a few hours, the analyst may lose the thread and needs help to recommence the work on the later occasion. Even where analytical work is continuous, some parts of the task will, inevitably, be put aside to concentrate elsewhere. The analyst is continually returning to points in the analysis which were previously left. It is vital, therefore, that progress is properly recorded to ensure that previous work is not misinterpreted.

Finally, it has been emphasised that goals, operations and plans cannot, in themselves, provide a comprehensive account of all that has been done. Part of the analysis process is concerned with justifying where the analysis has stopped. Reasons for stopping should be recorded, along with assumptions about the task context which have been made in deciding when to stop. Sometimes, a level of task detail may be felt to be sufficient because a training or other design suggestion has been made to support the appropriate operator behaviour. Unless these are recorded clearly, justifications made at the time will be forgotten. Hence, there will be insufficient guidance for people wishing to use the analysis for design and there will be no clear statement for anyone challenging assumptions made during the analysis.

The hierarchical diagram showing the relationship between goals, subgoals and plans has been the only method used to represent HTA so far and is generally preferred throughout the remainder of the book as the method for representing the task to illustrate various applications of HTA. However, diagrams are often insufficient for representing task detail in a practical project. It is usually advised to use tables to record detail and bring the main conclusions from both tables and diagrams to the attention of the client.

Hierarchical diagrams

Diagrams are very useful as a means of conveying the structure of the hierarchy, since the reader can easily trace various redescriptions. They are particularly helpful in navigating around the task, appreciating how different parts relate to each other and summarising the task for the client. However, they are unsatisfactory, on their own, as a method for communicating the various things that have been discovered in the analysis, such as task constraints, stopping decisions and design hypotheses.

Layout of diagrams

Layout of diagrams must obviously be clear. Vertical lines linking a goal to its redescription should be straight and easy for the reader to follow. It is generally helpful to keep as much of a task as possible on the same page, so that the reader is able to see how different parts of the task relate to each other. However, this can make for a very confusing diagram and prompts the analyst to use crooked lines to join goals to their redescriptions. In Figure 5.1 (a) some of the lines joining goals to their redescriptions have been bent (from goals 2 and 4). These have been straightened in Figure 5.1 (b) by modifying the layout of the diagram. Setting out a diagram on a single page without

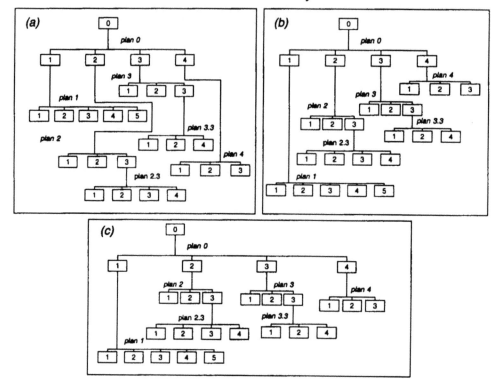

Figure 5.1   *Three versions of the same hierarchy showing different layout compromises.*

requiring the reader to trace lines through different angles can be complex. The analyst may choose to stretch the top horizontal line out to make space, as in Figure 5.1 (c), but this can cause the diagram to extend beyond the width of the page. The analyst may decide which parts of the task can be dropped to the lower parts of the page as in Figure 5.1 (b). Sometimes it is possible to tuck in part of a redescription in a convenient space, but it generally depends upon the size and complexity of the task and the graphics tools available for laying out the diagram

Including plans in the diagram

Plans can obviously cause clutter to a diagram, but they are an essential component of any analysis and cannot be missed out. Some people may argue that fixed sequence plans can be missed out provided their sequencing is preserved in the order in which they are set out across the page. This can avoid clutter, however, there is a risk of overemphasising the presence of fixed sequence plans. Analysts and their informants may too readily ascribe a plan as a fixed sequence and overlook an important contingency or choice that the operator must observe. They may simply record the plan as 'fixed sequence' without properly addressing its detail. It is generally reassuring to see a plan stated in full, because it confirms that the analyst has examined it carefully.

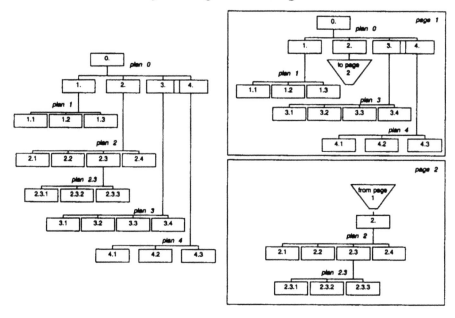

Figure 5.2   Representation of HTA over several pages.

Where plans are complex, they clutter diagrams. It is often helpful to record these more complex plans elsewhere and cross refer from the analysis diagram. Figure 3.5, which described the supermarket check-out operation, and Figure 3.20 in which the chlorine balancing task is represented, both show how plans are cross-referenced to a separate part of the diagram - to a separate box within the page or on another page.

Conventions for multiple pages

Where analyses get large, it is necessary to use multiple pages. This can apply during the process of analysis itself or when the analysis is finally presented. The important consideration here is to ensure continuity between pages. In this way, the analysis in Figure 5.2 can be represented on different pages. This process of linking pages is important both when presenting a full report and during analysis to avoid mixing up and losing work.

Showing where redescription has ceased

The analyst should signify to the reader where redescription has ceased. This can be done by underlining the box containing a goal where redescription is taken no further, as in Figure 5.3. I have not maintained this convention in most of the diagrams in the book in order to enhance the clarity of the diagrams, but it is a useful convention in real task analysis projects.

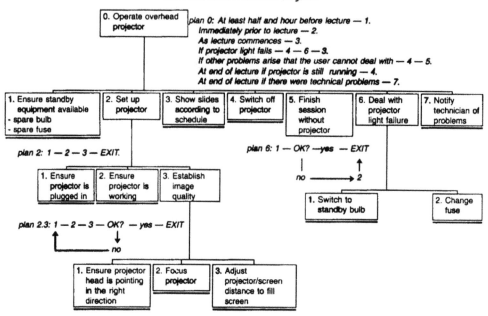

Figure 5.3   This shows both the manner in which goals not further redescribed are underlined and illustrates the numbering system.

Numbering the analysis

It is important to adopt a rational numbering system to keep track of the work. Numbering systems are merely clerical tools and different numbering systems can be devised. The system used throughout this book has proven to be effective. It is illustrated in Figure 5.3. The overall goal is given the number 0. Its subgoals are numbered from 1 to whatever is necessary — 7 in this case. A plan is given the same number as its superordinate goal (in this case 'plan 0'). The plan then makes reference to its subgoals by using the identifying number (e.g. 1 to 7). When operation 2 (Set up projector) is redescribed, its subgoals are again numbered from 1. The plan governing its 3 suboperations is labelled 'plan 2'. When operation 3 of 'Set up projector' is further redescribed, its subgoals are numbered 1 to 2 and its plan is given the label 'plan 2.3'. Thus, goals or operations are generally referred to by their single digit, while plans are always given a full number which shows their precise location within the hierarchy. This means that everything is easy to locate.

It would be possible to give the goals and operations their complete number within the hierarchy if one wished (e.g. referring to 'Focus projector' as operation 2.3.1). The reason for not doing this is that numbering becomes too cumbersome. In particular, reference to suboperations within each plan would take up too much space.

The reason for adopting the convention of numbering the top goal as zero is to facilitate subsequent revisions to the analysis. By referring to this as zero, then dropping reference to it thereafter, the analysis can be fitted easily into a wider task. Thus, if

'Operate overhead projector' were to become the fourth operation in a wider task involving the use of educational technology, then the only editing required would be to replace the zero with 4, then prefix each plan with the number '4'.

This numbering system is simple, effective and it works. It applies to diagrams and it also applies to tables, as will now be described. This enables cross referencing between these two forms of task representation.

Tabular Formats

The main objection to diagrams is that they do not easily and conveniently permit notes to be made concerning the task. Without backup notes, the task analysis is extremely limited. Tables provide a solution to this problem. There are two main issues concerning representing task analysis in a table. First there is the issue of how to represent the hierarchical structure in a linear sequence. Second is the question of what additional information should be provided in the analysis record. The first issue is concerned with the vertical layout of the table and the second is concerned with its horizontal design.

Sequencing the analysis

Numbering systems are most easily described by reference to an example. Figure 5.4 (a) shows the hierarchy of numbers relating to the supermarket task represented in Figure 3.5. To organise this into a sequence suitable for inclusion in a table, the overall goal is first recorded. Then its plan is recorded. Then its immediate subordinate operations are listed in order. Operations where there is no further redescription are signified by an appropriate symbol - here a forward slash (/) has been used. This process has developed the list shown in Figure 5.4 (b). The list is then inspected from the top to locate the first operation in the list that is further redescribed but which has, so far, not been considered. The only candidate here is subgoal 2. Subgoal 2 is now recorded as a goal awaiting redescription. Its subordinate operations and their plan are then recorded. The aggregate list, so far, is shown in Figure 5.4 (c). The process is repeated and the list grows to (d) and then (e) then (f).

Developing columns for the table

Two sorts of column need to be produced in order to develop an HTA table. The first type of column sets out the task description. The second type of column is used by the analyst to record comments and observations about the task. Table 5.1 represents the supermarket task set out in Figure 5.4 (d). In Table 5.1, the left-hand and centre columns show the task description, while the right-hand column illustrates how notes may be used to make observations. The left-hand column shows the goals, subgoals and plans. These are kept together so that the reader can easily see how plans relate to subgoals to meet the requirements of their superordinate. The centre column indicates where further redescription has ceased.

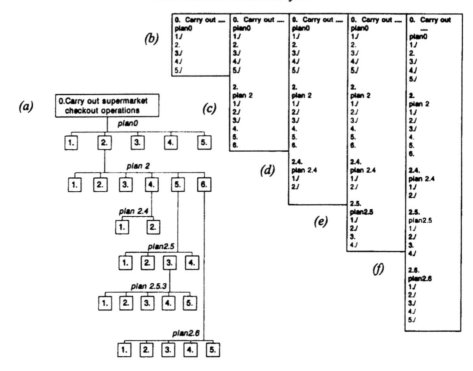

Figure 5.4   *The numerical representation of the supermarket checkout task from Figure 3.5 translated into a sequence for an HTA table.*

By inspecting the subgoals of plan 0, the next goal to be redescribed is 2, so this is copied to the next row and the process continues. When this is repeated, the next subgoal for consideration is subgoal 4 of goal 2. So this is copied to the next line. For ease of navigating through the table, however, the *full number* is now recorded for goal 2. (i.e. 2.4 rather than just 4). This means that wherever the reader now looks, he or she can navigate easily through the table. To find where 2.4 fits in, the reader traces up the table to locate Goal 2, then locates the 4th subgoal in its redescription. If the analyst is reading the redescription of goal 2 it is apparent that only subgoal 4 is further redescribed. So by tracing down the table and seeking the redescription of 2.4, the reader can locate the redescription required.

The right-hand column in Table 5.1 shows various notes that the analyst has recorded against the different task elements. 'Notes' columns may include notes of *explanation* about a task, *constraints* that the analyst and informant recognised about how the task is currently carried out, *hypotheses* about the kinds of performance weaknesses likely to be exposed and hypotheses for how the task might be improved, including suggestions to improve interfaces and develop training. Comments made in the 'notes' column of Table 5.1 are not particularly illuminating because the task is so straightforward. Table 5.2 shows the use of such a column in a task where there is greater cause for concern regarding how the task is carried out. This represents part of an analysis of a process control task in which the operator is exposed both to problems

Table 5.1   The main columns of the task analysis table.

Operations and Plans	Further redescription	Notes
0 Carry out supermarket checkout operations plan 0: At start of shift — 1. When customer presents — 2. At end of shift — 3. If there is a spillage on the conveyor — 4. If problems arise which cannot be dealt with — 5.		
1 Set till to start shift	no	Demonstration as per induction programme.
2 Deal with customers' purchases	yes	ditto
3 Complete till operations	no	ditto
4 Clean conveyor	no	ditto
5 Refer to supervisor	no	Note that supervisors are not always at their desk. Review methods of communication.
2 Deal with customers' purchases plan 2: 1 — 2 — if customer requests assistance — 3. Then — 4 — 5. If there is a spillage — 6. When customer has been dealt with, repeat from 1.		
1 Initialise till for next customer	no	
2 Establish whether customer requires assistance with packing	no	Some operators lack skill in enquiring of customers whether they require help.
3 Obtain assistance with packing	no	
4 Progress goods to till	yes	
5 Price individual items	yes	
6 Obtain payment	yes	
2.4. Progress goods to till plan 2.4 As necessary — 1 & 2.		
1 Remove customer divider	no	
2 Start conveyor	no	
2.5. Price individual items plan2.5: 1 — 2 — 3. If there are there further items, repeat from 1. Otherwise — 4 — EXIT		
1 Select next item	no	
2 Establish method of pricing	no	Training required for more rapid identification on non-bar-coded items.
3 Enter price	yes	
4 Establish total	no	
2.6. Obtain payment plan2.6: 1 — as selected by customer 2 to 5. If there is an authorisation problem repeat from 1.		
1 Establish method of payment	no	Training required to help operators distinguish between different types of card.
2 Deal with cash payment	no	More training required to identify forged notes and coins.
3 Deal with charge card payment	no	
4 Deal with credit card payment	no	
5 Deal with payment by cheque	no	
2.5.3 Enter price plan 2.5 See panel to left		
1 Weigh goods	no	
2 Enter price via weighed goods bar-code sheet	no	
3 Enter price via bar code	no	
4 Key in item code	no	
5 Enter price manually	no	

of lifting and carrying and to risks of toxic emission. In this example, the simple 'notes' column is used to enable the analyst to record a range of safety critical observations.

Other formats are possible which could lead to a more systematic examination. Thus Table 5.3 uses a simple taxonomy to prompt the analyst to record instances when planning activities could lead to risk, and then, for the ways in which performance may be affected. Planning issues might entail actions being carried out 'too early', 'too late' etc. Performance is examined using a simple taxonomy entailing 'perception', 'decision-making', 'planning', 'action' and 'feedback'. Thus, too much steam pressure during operation 1 could risk explosion; the aspect of performance that would cause the operator

Table 5.2   A simple tabular format showing the use of a 'notes' column. describing a range of human factors issues in a potentially hazardous environment.

Operations and Plans	Further redescribed	Notes
2 Make batch		
plan 2: 1 — after 20 minutes — 2 — 3 — 4 — 5 — 6 — 7. Throughout, as prompted — 8.		
1 Charge all materials into reactor	no	Batch formulation sheet needs to be reviewed.
2 Charge 1 measure of catalyst CB-48.	yes	
3 Heat to reflux	yes	
4 Reflux	yes	
5 Distil	yes	
6 Charge waxes, fines etc.	yes	
7 Ensure correct batch characteristics	yes	
8 Maintain batch record	yes	
2.2 Charge 1 measure of catalyst CB-48		
plan 2.2: In order 1 to 6.		
1 Obtain catalyst from store	no	Labelling conventions to be understood.
2 Don protective clothing	no	Safety precautions and consequences must be understood
3 Open man-lid	no	
4 Charge catalyst	no	Lifting and pouring in a difficult position. Manual handling training required. Ergonomics of tipping vessels should be reviewed.
5 Close man-lid	no	Operator needs to be clear that man-lid is secure. Should design of man-lid lock be reviewed to provide more positive confirmation?
6 Remove protective clothing	no	Is operator aware of consequences of removing clothing in the wrong order?

to exceed the recommended pressure is inadequate monitoring of feedback. The planning columns in Table 5.3 are based on ideas of *hazard and operability studies (HAZOP)* as applied to human error analysis (HEA) where the analyst systematically applies a set of key words to each operation identified, or at least, those operations which are considered to be of concern. All HEA methodologies require that a task is first analysed then systematically appraised by the analyst using an appropriate model of human error. The result is generally a table recording these judgements against different parts of the task. HTA has been use extensively within these methodologies.[25]

Table 5.3   Using columns in a task analysis table to allow the analyst to systematically reord comments about operations according to a given set of classifications.

	Planning					Operation			
	Too soon	Too late	Omitted	Too little	Too much	Perception	Decision	Planning	Action	Feedback
2.3 Heat to reflux										
plan 2.3: 1 — 2 when temperature is between 47° & 49° — 3 & 4 — when reactor level reaches below 0.25 mtrs — 5. If at any time reactor temperature exceeds 53° — 5										
1 Turn steam pressure to 6 psi. (no)					*				X
2 Monitor reactor temperatures intermittently until reflux is complete (no)	*	*				X			
3 Maintain distillate rate 40 cm per hour (no)	*							X	
4 Monitor reactor level (no)						X			

Representation of plans in diagrams and tables

Plans should be represented in a clear and concise way. Different analysts have their own preference for representing plans, as do clients and informants. Inspectors may also insist on particular formats to follow. Different alternatives are legitimate, provided the timings, sequences and conditions are properly reflected in the plan. There are many examples of plans in different formats throughout this book.

Location of plans

If both diagrams and tables are provided, a more complex plan may be omitted from the diagram by cross-referencing to the table where it appears in full. Even in tables, really complex plans may need to be attached as an appendix. In any event, the location of the plan must be clearly flagged at the point where the reader might usually expect to see it.

Representation in text

Some people prefer plans to be represented in text form. For example:

> *Do 1, then 2. When the temperature reaches 45°C, do 3.*

Lists

Lists are another textual form of representing plans. For example, the following list entails a number of operations carried out at the same time, with other operations contingent upon prior completion of other operations and external conditions. Where plans depend upon interpreting different conditions, this sort of representation can become cluttered.

> *START*
> *Do 1*
> *then together*
> > *{do 2, then when pressure > 10 psi do 3}*
> > *and*
> > *{do 4*
> > *If concentration outside tolerance, then*
> > *ABANDON OPERATION}*
> *If concentration is within tolerance AND operations 3 & 4*
> *are complete*
> > *do 5}*
> *EXIT*

Flow charts and rule-sets

If a situation is complicated and the procedure is contingent upon different circumstances, then flow charts and rule-sets are effective ways of expressing the plan. Figure 5.5

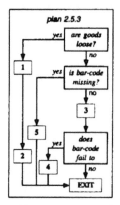

- *If goods are loose — 1 — 2.*

- *If goods are not loose, but the bar-code is missing — 5.*

- *If goods are not loose and bar-code is available — 3.*

- *If bar-code fails to register — 4.*

Figure 5.5   This compares a flow diagram with a rule set. This example is from the HTA described in Figure 3.5.

shows a flow diagram taken from the supermarket task, together with an equivalent text version. This text version sets out the decisions as a set of production rules ('if - then'). Sometimes it is best to favour the flow-diagram, for example where it helps an informant trace through the logic of the plan. The 'if-then' rule-set is sometimes preferred when the analyst wishes to capture a set of task conditions without necessarily making explicit the sequence of carrying out the task. The rules specify system conditions that must prevail to warrant suboperations being carried out. There may be several sequences of action that satisfy these rules and there may be no reason to constrain operators adopting whichever they favour.

Bar-charts/time-lines

Where there are complex time dependencies, it is often easier to use a time line, indicating the onset of actions and their cessation (see Figure 3.7, for example).

Computer aids in recording task analysis

The process of carrying out HTA, like all task analysis methods, generates copious tables and diagrams. There are, as yet, no wholly suitable computer tools to handle all of the aspects of handling and representing data for HTA[26]. There are, though, several features of common office applications that can be helpful. The following discussion is simply an introduction to possibilities. The reader must be left to explore his or her own computer tools.

Text outliners

Text outliners are available as parts of some wordprocessing applications and also as tools in their own right. They enable the analyst to type in the goals and plans and then organise them according to different levels. Outliners are sometimes called 'ideas processors' as they are sometimes promoted as tools to help organise thought. Figure 5.6 shows the outline of the supermarket task.

```
0.    Carry out supermarket checkout operations
      plan 0:        At    start    of    shift    —    1.      When    customer    presents    —    2.
      At end of shift — 3.  If there is a spillage on the conveyor — 4..  For any problems — 5.
      1.    Set till to start shift
      2.    Deal with customers' purchases
            plan 2:       see diagram
            1.            Initialise till for next customer
            2.            Establish whether customer requires assistance with packing
            3.            Obtain assistance with packing
            4.            Progress goods to till
                  plan 2.4:       As necessary — 1 & 2.
                  1.            Remove customer divider
                  2.            Start conveyor
            5.            Price individual items
                  plan 2.5:       1    —    2    —    3.        If    there    are    there    further    items,
                  repeat from 1.  Otherwise — 4 — EXIT
                  1.            Select next item
                  2.            Establish method of pricing
                  3.            Enter price
                        plan 2.5.3: See panel to left
                        1.            Weigh goods
                        2.            Enter price via weighed goods bar-code sheet
                        3.            Enter price via bar code
                        4.            Key in item code
                        5.            Enter price manually
                  4.            Establish total
            6.            Obtain payment
                  plan 2.6:       1    —    as    selected    by    customer    2    to    5.        If    there    is
                  an authorisation problem repeat from 1.
                  1.            Establish method of payment
                  2.            Deal with cash payment
                  3.            Deal with charge card payment
                  4.            Deal with credit card payment
                  5.            Deal with payment by cheque
      3.    Complete till operations
      4.    Clean conveyor
```

Figure 5.6   The outline of the supermarket task. This is taken from Microsoft Word 5.1. It is generated by 'demoting' those parts of the tasks that are subordinate to others.

Often, outliners provide a facility for the analyst to number parts of the task consistently. Another common feature of outliners is the ability to show only a chosen number of levels. For example, the analyst may only consider the top level, or choose to expand one of the operations to show more detail. The facility can allow the analyst to work directly with the outliner, typing in information whilst discussing the task with the informant. Information can be checked and rearranged before moving on.

Choice of format should depend upon clerical convenience and clarity. Using outliners is clerically convenient, and the result is clear when viewed on a computer screen, since the outline may be shrunk to view an individual redescription. However, it is easy to lose sight of the task structure when a full outline is printed out because different operations that relate to each other within the same plan become distant from one another. On balance the outliner is convenient, but beware the consequent problems of reading the printout!

Table facilities

Where a wordprocessor offers a table facility, it is a real boon. The table may be set out on the screen. The left-hand columns should contain the task analysis. The other columns can be extended as required.

Spreadsheets are even more helpful, because of their ease of formatting. Columns can be added at will, borders can be added to assist presentation, and it is sometimes possible to automate parts of the table-preparation process using macros provided within the spreadsheet. In earlier versions of spreadsheets the constraints on cell size limited the amount of text that could easily be recorded without having to add additional rows. Newer spreadsheets, however, have the facility to wrap round text and expand the depth of the cell within a defined cell width. This is ideal for creating task analysis tables. It is then possible to use the other spreadsheet functions, such as calculation functions in the case of quantification, or database functions to collect items warranting similar treatment. To create a table, the numbered outline from a wordprocessor can be rearranged and imported into the spreadsheet for additional formatting.

Text to diagrams

A further useful feature is the capability to convert a diagram to a table and vice versa. If the analyst is working in a diagram, it is helpful if the work can be automatically translated into table form where it can then be annotated and discussed. Some packages enable outlines to be converted into hierarchical diagrams, but these are rarely satisfactory for any but the simplest HTA. They are usually sold as methods to produce organisational charts showing the reporting relationships of personnel within companies. Unfortunately, the freedom to vary the size of text boxes or to rearrange layout to fit more onto a page means their use is usually limited for representing task analysis and they are unhelpful as tools used during the process of analysis.

Some effort has been made to automate the processes of diagram construction, so that diagrams can arrange themselves and tables be generated automatically. Many of the tasks engaged in sorting out a completed HTA require close attention to detail and can be cumbersome. Therefore, such developments are welcome. However, there is always the risk that this reduces the extent to which the analyst is engaged in the detail of the task. Sometimes, the very act of sorting out the clerical aspects of recording draws attention to problems of the accuracy of the task analysis. [27]

Concluding remarks

This chapter has set out a range of options for representing and recording HTAs. *Diagrams* are useful in showing how parts of the task relate to one another, although care needs to be exercised to ensure that the analysis can be followed. *Tables* are important to capture additional information related to the task. Both diagrams and tables preserve the hierarchical nature of the task description and generally complement each other as part of a task analysis report.

Chapter 6

Analysis of tasks - some illustrations

This chapter provides several illustrations of HTA to complement those already described. Examples have been selected to demonstrate analysis in a range of different task domains and to illustrate many of the elements of HTA. The tasks included are as follows.

- *Changing a cartridge in a printer or photocopier*
- *Operating a batch process plant*
- *Controlling a continuous process plant*
- *Air-traffic control*
- *Carrying out minimal access surgery*
- *Carrying out a customer service task*
- *Using a wordprocessor*
- *Carrying out mechanical maintenance*
- *Nursing*
- *Management*
- *Supervision*

Although the reader may select topics of interest, the examples are intended to provide a representative account of different applications of HTA. Where domains are not self-explanatory, a simple account of these domains is given. Note also that the examples given are mainly only extracts which have been selected to highlight distinctive features of the task.

Introduction

This chapter shows how HTA can be applied to different types of task. The examples discuss how and why the analysis was done, set out some of the significant aspects of the task analysis and indicate the key lessons to emerge from the analysis. The cases have been selected to demonstrate breadth of application and show how the same few principles of organising task information can be widely used. The detail of the task analyses serves to illustrate points made in the text and also to provide examples for people interested in specific domains. The tasks described are as follows.

Changing a cartridge in a printer or photocopier

This describes a straightforward procedure in which a series of actions and checks are linked together in order to replace toner cartridge in a laser printer. Procedures such as this are extremely common and occur in most task analyses.

Operating a batch process plant

Batch process plants manufacture materials by controlling physical and chemical changes to raw materials. Process control entails the operator making adjustments to process parameters in order to change processing conditions. A short explanation of process control is given.

Control tasks are extremely common and are encountered across a wide range of domains, including commercial and social contexts as well as industrial situations. Control tasks are often seen as associated with automated systems, but the concepts of control relate to any situation where the system under control has its own dynamics and may change even where the operator is not carrying out control actions. The operator's job is to study the system and decide what should be done to influence system behaviour.

Controlling a continuous process plant

Many systems rely on automation, leaving the operator to monitor the system, then intervene when things are not going to plan. Such tasks are seen in the supervision of automated transportation systems and in continuous process manufacturing plant. Skills entail monitoring and solving system problems. This example concerns supervision of a plant, of the type found in the petrochemical, food-processing and power industries.

Air-traffic control

There are several different air-traffic control tasks, those dealing with aircraft movement across air sectors and tasks concerned with takeoff and landing. The task described here deals with the activities involved in ensuring the safe movement of aircraft across an air sector. This example shows how different cognitive elements of a complex task interact.

Minimal access surgery

It is important to appreciate that the surgical skills involved in minimal access surgery are undertaken in the wider context of treating a patient. The surgeon must understand the patient's history and prognosis, since these can influence how the motor tasks are modified to suit the needs of the patient. This example is included to emphasise the importance of context in task analysis and to show how simple task elements become critical when understood in context.

A customer service task

Many tasks require operators to deal with the public over the telephone and use a computer database to order goods, manage accounts or provide other services. The presence of computers is sometimes given too much attention by analysts. The example will show how this type of activity must be understood in the context of the wider task.

Using a wordprocessor

The design and use of applications packages such as wordprocessors or spreadsheets poses some interesting challenges for task analysis. These computer programs are *tools* and their use needs to be understood in the context of the purpose to which they are being put. The example will show a strategy for understanding the use of this sort of computer tool.

Mechanical maintenance

Maintenance is critical to the successful performance of all systems. This example considers mechanical maintenance at a large industrial site. It deals with how individual maintenance *tasks* can be examined in the context of maintenance *jobs*.

Nursing in a hospital ward

Nursing involves a range of technical and interpersonal skills. Some activities are governed by the clock, while others are prompted according by events.

Management

Management is concerned with taking responsibility within an organisation to help meet an organisation's goals. The management task presented here is a general description of management duties, emphasising the importance of monitoring, planning and control in order to enable the system being managed to develop and function effectively.

Supervision

Where management may be concerned with setting up systems, supervision is concerned with making sure an existing system works properly. In many respects, supervisory tasks are like process control tasks. This example deals with a supervisory task relating to nursing on a hospital ward.

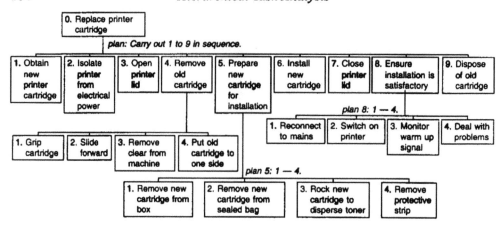

Figure 6.1   Changing a cartridge in a laser printer.

Changing a cartridge

Changing a cartridge in a laser printer or a photocopier is a relatively common office routine (see Figure 6.1). It entails doing things in a series of steps, usually in accordance with clearly written instructions, until the printer or photocopier is back into operation. It is similar in structure to many maintenance tasks and many assembly tasks. Sometimes the constituent steps require skill, but often this skill can be minimised by effective equipment design. The design of most printers means that inserting the cartridge the wrong way round is impossible; only when the cartridge is correctly inserted can the lid be closed and the printer switched on.. In this way, the operator gains rapid feedback concerning the success or otherwise of the action.

Process control

In situations such as changing a printer cartridge the operator physically manipulates objects in order to accomplish the task. Left alone, the printer and the cartridge would lie on the desk and do nothing. In contrast, there are many situations where the systems being controlled have their own dynamics, especially where automation is involved. In *process control tasks* it is the operator's job to manipulate controls to achieve a set of conditions in the system such that the system does the things that it is supposed to do.

The term process plant refers to industrial plant where operating conditions are adjusted in order that feedstock can be subjected to controlled physical conditions to enable certain physical processes to occur. For example, in the manufacture of dried milk, fresh milk is sprayed into a tower where it is subjected to heat which causes it to dry. The spray must be adjusted to ensure the particles are of a standard size and the temperature must be carefully controlled to ensure the correct rate of drying - too low would cause the particles to collect into a wet lump; too high would cause burning,

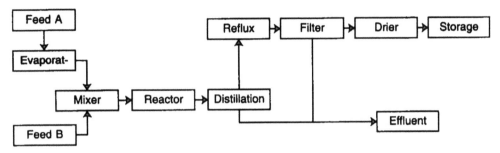

Figure 6.2   A typical arrangement of processing units in a process plant.

resulting in ruined batches and the risk of fire. Similarly, in the manufacture of antibiotics, raw materials must be kept at a suitable condition to enable microorganisms to grow. Process control is used in a variety of situations. As well as food-processing and pharmaceuticals, process control is prominent in the chemical, petrochemical, power generation and brewing industries. Principal characteristics are that the operator manipulates controls to change parameters and often depends upon instrumentation to make judgements - for reasons of safety, productivity and hygiene.

Process plants are made up from a number of different stages called *unit operations*. These stages may mix, heat, cool, distil, dry, filter, or store materials. Which of these unit operations are used for a particular process depends upon how the plant designer decides to process raw materials in order to obtain the desired product. Figure 6.2 shows a possible arrangement of unit operations. These are not necessarily separate vessels, but rather, separate stages in transforming raw materials to finished product.

There are two main types of manufacturing in the process industries - *batch* processing and *continuous* processing. *Batch* processing relies on a specified amount of raw materials being introduced as feed, then being moved through the unit operations step-by-step. For several steps the batch operator may not need physically to transfer materials between vessels. For other steps, the operator may be engaged in transfer between vessels - for example in pumping, conveying or even shovelling. An implication for the process operator is that attention moves from handling the introduction of the raw materials, through managing the evaporation, then managing the reaction, and so on. Depending on design, some plants require operators to engage in cycles of manual handling, then monitoring, then diagnosis, then laboratory analysis. Sometimes the operator is also required to start the next batch when the previous batch is clear from a vessel. Therefore, the operator is continually engaged with a number of contrasting tasks and must often learn how to time-share to maximise productivity. A further characteristic is that batch processing often requires the operators to work close to vessels and to product, even opening vessels to make inspection. These activities are often hazardous, so careful attention to safety and to correct procedures is crucial.

In *continuous* processes, raw materials are introduced into plant in a continuous stream with a view to maintaining a continuous output of product. Petroleum refineries are continuous processes because there is a continuous consumption of petroleum products throughout the world. Continued sales are virtually guaranteed, so continuous production is essential to maintain stocks. The same applies to power generation. Gas and electricity are in constant demand, so a continuous feed of raw material is maintained to achieve a continuous output for the national grids. This continuous demand prompts continuous processing. A continuous plant is arranged so that all unit operations are being maintained at the same time. This means that during full production, temperatures, pressures, levels and flows, as well as other critical parameters, are maintained so that the materials flowing through are always subject to the correct conditions to enable processing to take place.

The plant represented in Figure 6.2 could be a batch plant where each batch of new raw material is subjected to each unit in turn. Alternatively, it could be a continuous process where all units are operating simultaneously, with product flowing through continually. The continuous process plant operator is subject to different pressures to the batch plant operator. To maintain conditions in a continuous process plant, most aspects of the control system are automated. Moreover, to maximise profit, plants are large. It means that the operator's job is mainly to supervise a large, often hazardous, automated system. While the batch plant operator is often busy, the continuous plant operator may be subject to periods when little of concern is happening - because the automatic control system is doing its job. Nonetheless, the continuous process plant operator must continually monitor plant conditions to ensure that they are within acceptable tolerances to ensure safety and productivity. Then, if something does happen to cause a disturbance, the operator must respond quickly to work out what to do to make the system safe, then manage its recovery to full production again. Despite the fact that these plants can be distinguished according to their production patterns, there are many overlaps in skills. Continuous process plant operators are often engaged in managing procedures, especially when plants are starting up and shutting down. Equally, batch plant operators are often engaged in periods when they must maintain the batch at a steady state, to enable a reaction to take place.[28] The case of the batch plant operator will be dealt with first.

Task analysis of a batch operation

While the analysis of a batch process control task is guided to some extent by the steps that are followed in the process, this is not where analysis should start. Figure 6.3 shows the top level of the analysis of a typical batch process control task. This reflects the overall cycle of the job that people have to do. Operations 1 and 8 are handover operations concerned with the need to maintain continuity from shift to shift. Operation 3 is concerned with making sure opportunities are taken to get the next batch underway without delay. Operations 6 and 7 are concerned with unscheduled events that occur during production. Operation 4 (maintain production) focus specifically on the unit operations to be dealt with to process the batch. It is important to capture this overall level of any analysis, because it identifies skills concerned with managing the job and not simply the manufacturing process.

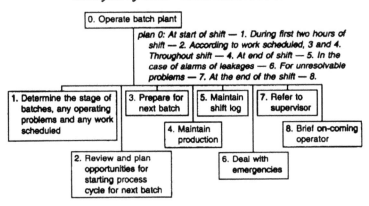

Figure 6.3   The top level of an HTA for a batch plant operation.

Turning to the main processing steps of the batch, these emerge from the analysis of operation 4 and are shown in Figure 6.4. Goals 2 to 11 in Figure 6.4 reflect the stages in chemical processing, but since this is shift-work, and since more than one batch may be involved, it is inappropriate to start at the beginning and work through. Therefore, the plan requires the operator to monitor conditions throughout to see what to do next. The cue to the next action is contained within the plant procedures, often called 'Manufacturing Instructions' - which we will return to in Chapter 11. Thus, the operator must, through monitoring, be aware of what steps are required next and be alert to the conditions - time and batch progress - when these steps should be carried out. Each of these steps in batch production are developed further. Operation 6 in Figure 6.4 illustrates

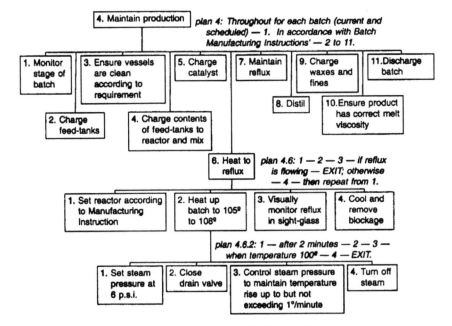

Figure 6.4   The manufacturing phase of the batch plant task analysis.

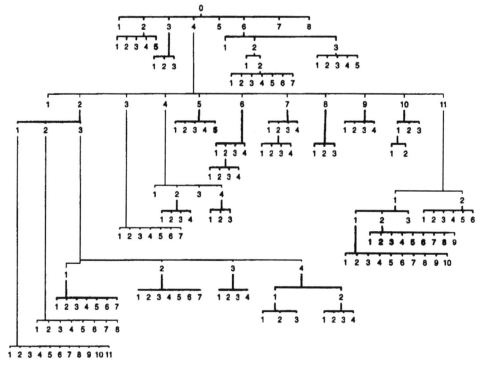

Figure 6.5   This extent of the HTA developed for the batch processing task.

how one of these steps - heat to reflux - is carried out. Typically, this involves monitoring information, making judgements and carrying out actions. The plans contain contingent sequences and cycles, typical of this type of work.

It is impossible to present the whole of this task analysis here. However, Figure 6.5 shows the extent of the full HTA that was developed for this project. It represents the hierarchy of goals but not the plans. This is by no means a large analysis, although it may look complicated and daunting. Conducting an analysis of this magnitude requires care and attention, but it is not necessarily difficult. The extracts from this analysis (in Figures 6.3 and 6.4) show that each redescription is reasonably straightforward. The complexity emerges by combining many of these straightforward plans. The *size* of the analysis is dictated by the requirements of the task. The rules and methods of HTA mean that this kind of complexity can be handled with relative ease, provided an appropriate method of recording task analysis is used (see Chapter 5).

By making reference to 'manufacturing instructions (MIs)' in plans, the task can more easily be adapted to enable the operator to deal with other products by using different MIs. Thus, a common set of skills and procedures are adapted in order to produce different products.

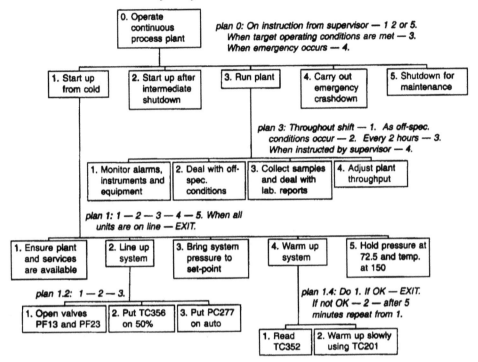

Figure 6.6   Extracts from an HTA of a typical continuous process control task.

Task analysis of a continuous process control task

Whereas batch processing requires operators to engage in changing conditions throughout production, continuous processes need far less intervention if they are working properly. Industries such as oil refining or power generation rely on large complex automated plants. As described above, the plant represented in the block diagram in Figure 6.2 could be a continuous process plant in which the operator would try to maintain appropriate operating conditions in each of the vessels at the same time.

Figure 6.6 shows extracts from the HTA of a continuous process control task. The first redescription contains subgoals concerned with starting up the plant, running the plant and shutting the plant down. Note, this analysis represents the control task and not the job the operator is employed to do. To look at a continuous process plant operator's job would entail a top level such as that in Figure 6.3, because, as in all process industries, staff hand over responsibilities to other staff when a shift has finished in order to maintain the ongoing process.

Depending on the system, there could be 'start-up from cold and empty' and 'start-up from an intermediate state'. Following a problem, the plant might be held in an intermediate state while maintenance is carried out. When target operating conditions are to be recovered, the vessels may be still warm and partially full. Therefore, the strategy for recovering from different intermediate states may vary. 'Start-up from cold' may entail a long, invariant, contingent procedure, whereas starting up from an intermediate state may entail more ingenuity and planning on the part of the operator. Often, 'start-up' is best seen in terms of the operator attaining intermediate states, rather than simply treating the procedure as a set of actions. So, plan 1 may be treated as an invariant procedure because it shows that important intermediate states of the system must be achieved in order. However, problems may occur during each of these stages that the operator must deal with before moving on - recall the analysis of the distillation training in Figure 3.8. In this way, the operator might be engaged in a diagnostic and recovery procedure that we would normally anticipate during plant running (operation 3.2 in Figure 6.6 and further expanded in Figure 6.7).

Equally shutdown could be an emergency shutdown or it could be controlled. If a problem arose which needed dealing with instantly in order to avert a crisis, then most systems have an automatic procedure, which quickly moves the system to a safe state. However, starting again from these safe states may be time consuming. If the operator was able to control the shutdown, it may be possible to restart the plant in a more efficient way. Emergency shutdowns need to be as prescribed as possible in order quickly to achieve a safe state. Controlled shutdowns may need the operator to carry out more planning. Thus, emergency shutdowns may warrant frequent routine drills and possibly the use of job-aids, whereas, to ensure that controlled shutdowns are conducted safely and efficiently will require planning skills that may need to be practised in a simulator. Through the task analysis process, the client can indicate which modes of shutdown are required. These will have different requirements for supporting performance, so it is important that they are understood by the analyst.

When target operating conditions have been achieved following start-up, the operator can be said to be in the plant running phase - operation 3 in Figure 6.6. Running a plant entails a number of routine operations, such as collecting samples. There can be operations concerned with changing operating conditions to make product to a different specification. These may be achieved by following standard procedures or they may entail planning and decision-making. Monitoring behaviour, as in operation 3.1 in Figure 6.6, entails paying proper attention, especially to those things currently of concern. Many of these systems will be alarmed - flashing lights or auditory alarms will signal that the plant has gone off specification. In many cases, however, there is still reliance on the conscientious operator keeping an eye on progress to detect when conditions are

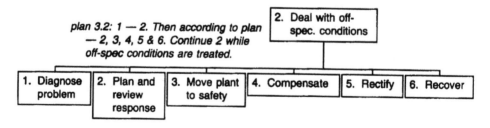

Figure 6.7   Redescription of dealing with off-specification conditions in a continuous process control task.

not as they should be. If a parameter is outside of specification, the operator should first decide whether this is of real concern or whether it could be a temporary situation that will correct itself. If, following this examination, there is cause for concern, the operator must deal with off-specification conditions.

Figure 6.7 shows how dealing with off-specification conditions might be analysed. Depending on the technology and the risk, the operator might be given more or less discretion. In some situations the operator would be required to diagnose and plan a response by reading an operating instruction. In other situations, the operator might be expected to reason and devise the optimal way of dealing with the problem. Elsewhere, the operator might be required to refer the problem to a more qualified colleague. Some situations enable the operator to compensate for the problem to optimise production while the main problem is resolved. In other situations, the operator must move to a safe state before doing anything else. The operator will be required, at some stage, to rectify the situation, either directly or by recruiting the support of maintenance staff. When the plant is fully operational again, the operator must manage the recovery of the process.

It is important to note, with respect to Figure 6.7, that the HTA has set out what people should do and not described how people actually do these things - that is, it is their responsibility to diagnose and plan responses, move the plant to safety, compensate for disturbances, etc. Some of these operations can be examined further using HTA, but if the operator is required to use expertise to devise novel solutions to unforeseen problems then the analyst might use other approaches to explore this behaviour further. As discussed in Chapter 4, this might entail undertaking a *cognitive task analysis*, although often such detail is unnecessary. For example, people can master skills such as diagnosis and planning, by being provided with scenarios in which they must diagnose and plan. To provide such training, simple scenarios can be initially chosen for practise using an appropriate form of simulation to enable the trainee to apply the skills and knowledge which they have been taught. Then the scenarios chosen for practise can be made more difficult. Training can continue until the trainee is performing satisfactorily.

Air-traffic control

The term *process control* is usually applied to manufacturing systems, but many of the same considerations apply in other contexts where the operator is employed to ensure that a system with its own dynamics performs according to expectation or requirement. The *Air-traffic controller's task* is to ensure that aircraft proceed through airspace in a safe manner. For most flights, flight plans are prepared which then generate information that can be used by the air-traffic controller to anticipate when aircraft are due into the air sector they are controlling, from where they are coming and where they are planning to go. This information will also indicate destination and preferred flight path. The air-traffic controller's task is to judge whether conflicts may occur between the flight paths of individual aircraft and then take steps to resolve these conflicts by requiring aircraft to climb or descend to different levels. (This is an oversimplification, because there are

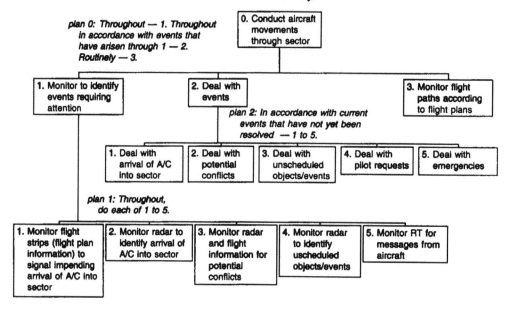

Figure 6.8   The air-traffic control task.

many factors that need to be considered and there are many variants on this basic task - but the features described here are characteristic.)[29]

A full analysis of these tasks reveals a range of procedures that must be followed, terminology that must be used, knowledge that controllers need in order to make decisions and skills they need to plan and make adjustments to flight plans. The extract shown in Figure 6.8 is only part of the overall job cycle. Air-traffic controllers work in short sessions of about one hour to an hour and a half. These periods are limited to prevent controllers becoming fatigued as a result of the intense monitoring required. Each time they assume responsibility for the task they must adapt to the previous controller's strategy. Then they must hand over to a colleague when their session is complete. Figure 6.8 shows general goals of 'monitoring to identify events requiring attention', 'dealing with these events' and 'monitoring flight paths according to flight plans'. Even with this small part of the task analysis, a number of implications can be drawn. First of all, there are two sorts of monitoring. One sort (operation 1) is concerned with detection - the controller must identify, as quickly as possible, when certain events arise. The other sort (operation 3) is concerned with maintaining conditions - the controller must keep an eye on each aircraft to determine whether it is operating as expected and deal with any variations. A second main implication is that, according to the plan, the operator must do 1 continually (which really means, as frequently as possible), whereas operation 2 is only required as events arise that need to be dealt with. Operation 3, concerned with maintaining the aircraft on its route, is done intermittently, but reasonably frequently to ensure that the aircraft is where it is supposed to be. When we consider the manner in which events can arise, it is clear that the operator is subject to substantial workload whenever several events occur together.

Moreover, while attention is directed towards dealing with these events, the controller still needs to monitor the system to detect other events that may require attention and to ensure that each aircraft in the sector is making satisfactory progress. The separate decisions and procedures of the Air-traffic controller's task each require considerable skill in their own right, but perhaps the most significant aspect of this sort of work is that the controller may need to deal with many such events at the same time according to what happens during a session.

This task is clearly heavily cognitive. It is concerned primarily with perception, monitoring detection, planning and other sorts of decision-making. But it does not immediately follow that we need to go to great lengths in understanding this cognition. Operation 1 is concerned with different sorts of monitoring. While these things are all done 'throughout' according to the plan, the experienced controller will do each as necessary in accordance with patterns of expectation. For example, operation 1.1 (Monitor radar to identify arrival of A/C into sector) will start to be done when the controller has already seen, from operation 1.1, that a new aircraft is expected arrive shortly. Equally, the experienced controller will focus on potential conflicts (operation 1.3) more closely on some routes and at certain times of the day than other.

Minimal Access Surgery (MAS)

While changing a cartridge is a relatively straightforward task which most people, given written instructions could undertake, minimal access or keyhole surgery is clearly something carried out by people with special expertise. It embodies many of the properties of a procedure - a distinct start and end point, with intervening goals that are carried out in sequence, even though there is great flexibility with regard to how these goals themselves are carried out. Minimal access surgery entails a set of techniques that are applied to a wide range of operations. If we simply think in terms of how minimal access surgery is different to other forms of surgery and focus on the manipulation of instruments during the operation, we may miss some key factors.

In a task analysis like this, it is important to step back to appreciate the wider task context, because keyhole surgery is just part of a wider pattern of treatment that a patient will receive. The patient first presents symptoms to the medical system. Before keyhole surgery is undertaken, a doctor will assess the patient in order to consider which treatments, if any, are to follow. Considerations at this stage entail clinical judgements and also judgements about prognosis and the consequences of different outcomes for the patient. These considerations can affect which courses of treatment will be followed and will also affect how different stages in the treatment will be executed. In the case of dealing with ectopic pregnancies, for example, knowledge of the history and circumstances of the pregnant woman is important. For example, if it is known that the patient has no children, then different decisions might be made during surgery where a pregnancy had not been planned and where the mother had already expressed anxiety about a further birth. These issues are concerned with surgeons identifying and weighing risks. Equally, knowledge of prior operations on a knee and

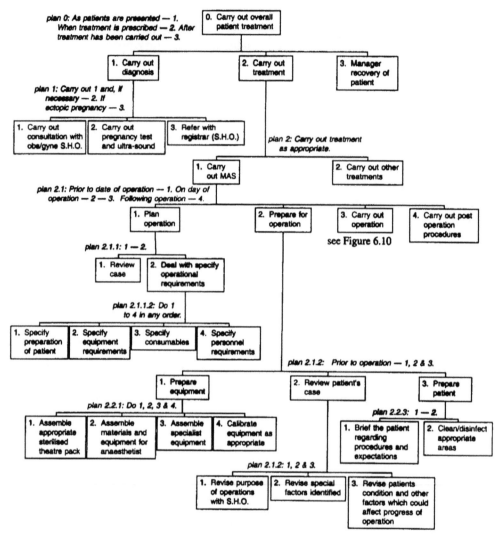

Figure 6.9   The minimal access surgery task.

knowledge of whether the person plays sport requiring certain kinds of movement, can affect decisions taken as an operation progresses. As the operation progresses, the surgeon makes judgements about the state of tissue being observed during the operation and must vary strategy and in some cases modify the purpose of the operation. Damage to a knee might be seen as more drastic than had been anticipated and so the operation may be modified or aborted. In the same way, knowledge of how an operation has progressed can influence the decisions that the consultant will subsequently make in making recommendations for courses of remedial therapy, carried out by a physiotherapist, for example.

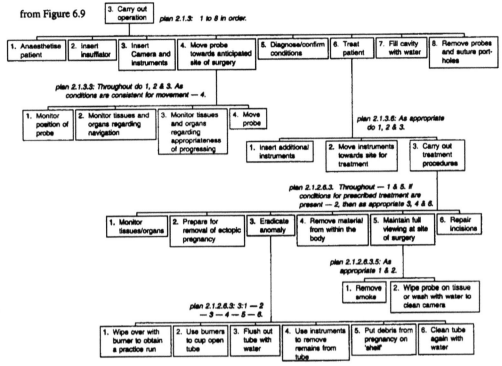

Figure 6.10 Detail from the minimal access surgery task dealing with a number of psychomotor skills.

With these wider considerations, it is clear that the surgeon's skills and decision-making are influenced by what went before and what might follow. If these things are not understood by the analyst, their influence on the decisions taken during surgery cannot be taken into consideration. Thus, the HTA of the keyhole surgery task most usefully begins by considering the full cycle of treatment, as shown in Figure 6.9. This describes a series of stages commencing from where the patient is first presented to the surgeon for examination right through to where the surgeon specifies a pattern of recuperation. Within this description, operation 3 deals with the specific task of keyhole surgery. This part is expanded in Figure 6.10 which describes the surgery involved in dealing with an ectopic pregnancy.

Minimal access surgery can be applied to a number of operations which follow similar patterns using common techniques, but which vary in detail. For example, operations on internal organs require some different procedures to operations on limbs, because the cavities within which the surgeon works have different physical characteristics and require different techniques to enable movement of instruments. If an analyst was required to consider keyhole surgery in general, this is best done by first focusing on a specific type of surgery, then, when that analysis is complete, the next type is considered. The analyst will now find that several parts of the task are common

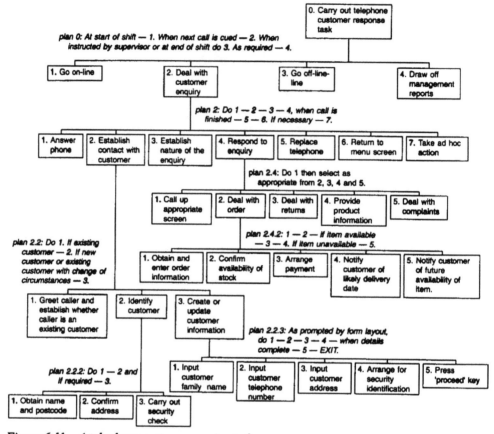

Figure 6.11   A telephone customer service task.

to the earlier example; only those parts that are different need to be considered. Then another type of surgery will yield fewer 'new' aspects. Soon, a number of different sorts of operation will be understood by reference to a common model in which many aspects of the tasks require similar skills.

A customer service task

Appropriate use of computers is critical to the success of many organisations. Increasingly, organisations make contact with their customers over the telephone and use telesales operators to identify customer needs and help fulfil these needs by interacting with the organisation's computer. Similar tasks exist in the services industries, such as providing gas and electricity services to domestic customers. There is nothing particularly special about tasks involving computers, indeed, they are often more straightforward than most because computer interfaces are often reasonably well designed and have been specifically tailored to the user's needs. A good interface will prompt the user concerning actions that will achieve the user's goals. This minimises the need to learn complicated sequences of action, hence, training is minimised. Should the user press the wrong key, an 'undo' function will often return the computer to its previous state, so any error should be easy to detect and rectify. Where technology and the task prevent the designer from developing a good interface, the task will be more difficult and more training will be required to overcome these difficulties.

The telephone customer service task in Figure 6.11 shows a sequence of activity commencing with the procedures of switching on equipment at the start of shift and finishing with the task of drawing off management reports. Operation 2 focuses on dealing with customer enquiries. This is a sequence of activity that is repeated each time the phone rings. A crucial early action is to identify the caller. This may entail obtaining the customer's name and postcode or a customer identification number. Sometimes the telephone system aids this identification process by automatically identifying the caller's location. Generally, as indentifiers are entered into the computer, further details of the customer will appear, prompting the operator to request confirmation. Different calls warrant different responses and these are reflected in goal 2.4.

The significant feature of this task analysis is that it examines the *whole* task, of which the computer interaction is only one part. The operator is the interface between the customer and the organisation; the operator's task is to represent the customer to the organisation and the organisation to the customer. The flexibility that a human operator can bring is often essential because customers may be unclear about what they require. Good interface and task design is especially important, therefore, because operators need to focus on the customer's needs and not allow dealing with the computer interface to interrupt this discourse.

Using a wordprocessor

Wordprocessors, spreadsheets and design program's, for example, are computer applications which present the user with a set of *functions* which enable them to carry out certain tasks. Such programs are *tools*, in the same way that a hammer or an egg-whisk is a tool. Used appropriately, they provide considerable assistance to the user. Unused, they lay dormant. We cannot carry out a task analysis of a wordprocessor, but we can analyse tasks in which wordprocessors are used. This can provide useful insight about the nature of the program itself. An analyst might wish to undertake the task analysis of a wordprocessing package to evaluate the interface, compare how two or more wordprocessors are similar or different, develop a user support document or help · prescribe training. A first consideration is that, while there are many generic skills in wordprocessing, different programs work in different ways. Therefore, any task analysis should focus on a specific program, even though the results of several might then be compared to identify core skills in order to provide elements of a wordprocessing course, for example.

The strategy for examining the use of the wordprocessor using HTA is to devise a series of tasks which, taken together, will account for the range of functions of interest.

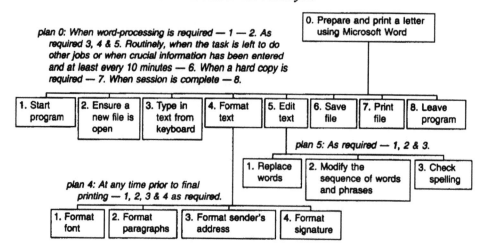

Figure 6.12   *Preparing and printing a letter in a wordprocessor.*

For example, by choosing something straightforward, such as 'writing a letter', the analyst will encounter the basic processes of starting and quitting the program, entering text, simple formatting and printing. Following this, the analyst can consider something more complex, such as 'writing a short report', which will involve continuing to practise existing skills plus adding new ones, such as, numbering pages and creating contents pages.

Figure 6.12 shows part of the analysis of the task of preparing and printing a letter using Microsoft Word™ . This is the higher level of description of the task of writing a letter. It covers the main activities. Many plans refer to the user's choices, because that

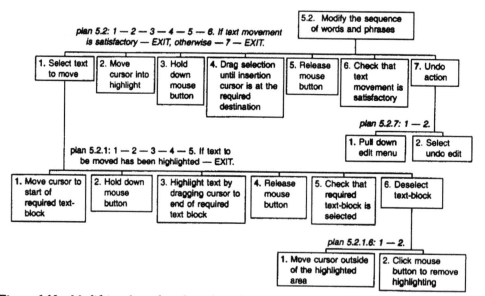

Figure 6.13   *Modifying the order of words in the text.*

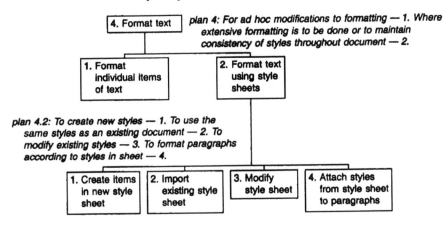

Figure 6.14   Ways of developing text formatting.

is what using a tool requires. Writing a letter relies on the user knowing how to compose a letter and what the content and length of the letter will be.

Figure 6.13 considers one of the subordinate goals that might be developed further. This example is concerned with modifying the order of words and phrases in order to rephrase text and make corrections. Here, the analysis can continue until operations are expressed as actions within the wordprocessor. It is noticeable that this level of detail starts to reveal common routines that are applied throughout the use of the program, for example, how to undo an unwanted action (operation 5.2.7) is revealed. Equally, procedures for highlighting, used here for moving text blocks (operation 5.2.1) are the same as when blocks of text are to be reformatted. These are procedures that can be picked up later on, saving time for the analyst.

Further tasks can be analysed that encompass additional functions of the wordprocessor. These further analyses can be used to modify the earlier versions. Thus, Figure 6.14 shows how more advanced wordprocessing skills can be incorporated. These functions will vary between products, although many such features are shared. An example is the use of style sheets for text formatting. Using a style sheet entails generating standard properties for blocks of text, then applying these as needed. For example, a user might decide that where a quotation is included in a document it will always be presented in italics, inset by 1 cm and with extra space above and below. In this way, a single style called 'quote' is created and can then be applied easily and consistently throughout the document. Moreover, if the user then decides that the format of quotations does not look very good after all and prefers, instead, that they are underlined and indented by 2 cm, then the 'quote' style can be edited. This will result in all blocks of text to which the 'quote' style is attached changing to the new format. To extend the HTA in this way the analyst needs to find out what these advanced variations might be and then look for text editing tasks where they are used. As the process develops, further wordprocessing concepts that the user needs are identified.

The plans that emerge from such analyses of computer applications packages generally conform to procedures or choices. *Procedures* are the steps that the user must follow to achieve a required outcome. So these are things that must be learned, or prompted by effective interface design or followed in a manual. *Choices* are usually reflections of the user's preferences - what material to enter, how to present it, etc.

Depending on the analyst's purpose, the HTA can progress in different ways to reveal different things. The analyst might choose to examine all procedures in detail to determine whether the design of methods are consistent or economical throughout the program. If the purpose of the analysis is to support training or the development of a user guide, then examination of plans will reveal where choices are made. This will point to the concepts the user will need in order to make these choices.

Mechanical maintenance

The principal purpose of *maintenance* is to ensure that systems are kept in a suitable state to ensure that they can be operated to fulfil the purpose for which they were

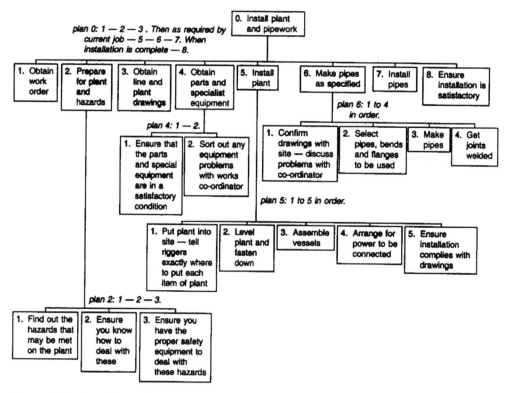

Figure 6.15   Installing plant and pipework — illustration of a typical mechanical maintenance task.

designed. Thus maintenance might be concerned with *preventing* problems or quickly *dealing* with them when they arise. When we encounter maintenance tasks in industry we see a mixture of activities which can involve the installation of new or replacement equipment, diagnosis and repair of equipment *in situ* due to some failure of the overall system, or removal of equipment and subsequent maintenance and repair in a workshop. Changing the cartridge in a printer or photocopier is a maintenance task, because it is designed to ensure that a piece of equipment again performs to the required standard. Equally, the ultrasonic testing task and the systems engineering tasks, described in Chapter 4 were types of maintenance task.

Mechanical maintenance tasks, such as repairing items of equipment, are parts of maintenance jobs and operate within maintenance systems. One difficulty in analysing maintenance tasks is capturing the way in which the same sorts of procedures must be applied to many different pieces of equipment. The analyst's problem is to work out how these variations can be accommodated sensibly within the single analysis. We have already seen two instances of how this can be done using HTA. In the case of minimal access surgery, the analyst first focused on a single surgical operation, then considered a second surgical operation to establish how the second differed. This process lead to recognising where tasks are similar and different - the general engagement with the patient was similar across surgical operations, but the detail of certain tasks varied between operations. Similarly, in the case of the wordprocessor analysis, focus upon the simple task of writing a letter revealed a general task structure that could be applied to more complex tasks. The same strategy is applied when considering the analysis of a multi-batch process plant used to manufacture products to different specifications using the same equipment.

To examine tasks such as maintenance it is helpful to focus first on a single task activity. Figure 6.15 shows the task analysis that emerged from considering one common maintenance activity, namely, installing plant and pipework[30]. The task analysis shows how the maintenance fitter must first obtain the relevant paperwork, must become familiar with safety issues, must obtain tools and equipment and must then carry out various craft tasks to install the equipment - indeed, mechanical maintenance is not simply confined to exercising mechanical craft skills.

There are two ways in which this HTA needs to be extended. First, the technician must do more than *instal* new equipment. Other types of activity include *dealing with faults, carrying out scheduled maintenance* and *maintaining in the workshop*. Second, the HTA must be extended to cope with the other items of equipment and plant subsystems that need to be maintained. For example, the technician needs to cope with *pumps, valves, centrifuges, refrigeration units* and many other items.

To achieve the first of these a general analysis of the mechanical maintenance technician's task was sought. When the HTA of *install plant and pipework* is examined, it is apparent that some activities are concerned with general steps in engaging with a maintenance problem, while other activities focus on a particular context. It is possible to represent the overall task of the maintenance technician's task as something like that shown in Figure 6.16. This summarises the range of things that are in the technician's job-description.

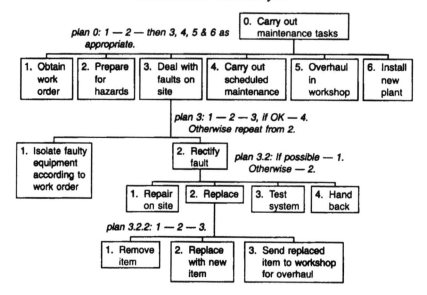

Figure 6.16   Representation of the general maintenance job

With regard to the second problem - the range of plant and equipment to be dealt with by maintenance technicians - an inventory was taken of all the equipment on the site. Table 6.1 shows how this inventory was recorded. Different items of equipment were listed to define *rows* in a matrix (in the real application, different types of pump and valve were also distinguished). Then the different sorts of maintenance activity were set out to define the *columns* in the matrix. Then each cell was considered, in turn to decide whether analysis was warranted. The shaded cells indicated where the task did not exist, for example refrigeration systems were never removed to the workshop for repair and dealing with faults did not apply to pipework. The remaining cells represented sub-tasks that required attention. This did not mean that every combination needed detailed analysis, because as the analysis of later sub-tasks was undertaken, it became apparent that much detail had already been completed already when considering earlier tasks. It was also possible simply to focus on the specific task undertaken in context. For example, in the analysis in Figure 6.17 it was possible to focus on just the

Table 6.1   Table of sub-tasks to be analysed for the mechanical maintenance project.

		A	B	C	D
		Install plant	Deal with faults	Scheduled maintenance	Workshop repair
1	Valves				
2	Pumps				
3	Pipework				
4	Weigh scales				
5	Refrigeration system				
	etc.				

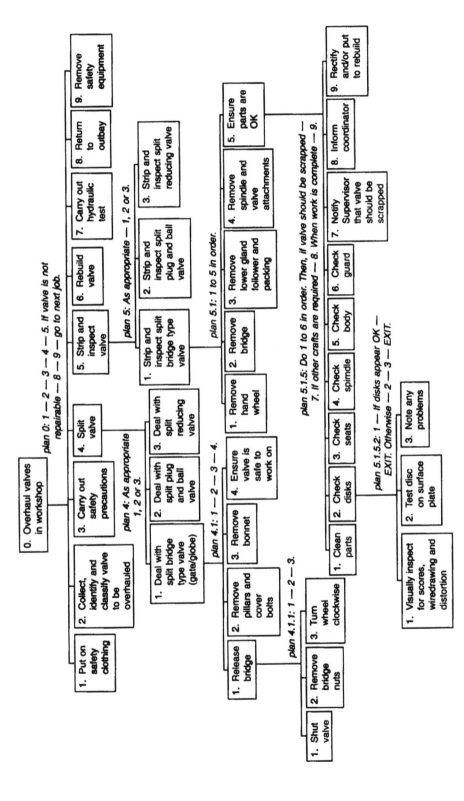

Figure 6.17 Analysis of 'overhaul valves in workshop which focuses on maintenance of a specific type of equipment in a particular domain.

activities that were concerned with *overhauling the valve in the workshop*. Some of the skills relevant to this were also encountered in other workshop tasks and some skills were also relevant to carrying out repairs in the workshop on other items of equipment.

Despite these methods of economising on analysis, this project required a substantial effort for the analyst and the company. Such projects are only worth undertaking if the outcome is important. Even then, it is advisable to seek as many benefits from this sort of exercise as possible. This work could provide an audit to check that tasks were safe, it could provide the basis for developing procedural guides and for developing and administering apprentice training.

Nursing

Many of the tasks discussed so far have been concerned with industrial or commercial applications. This next example deals with the duties of a nurse working in a hospital ward. Nursing duties are extremely varied and entail mixing rigid procedures with discretion. They also entail diagnostic skills, where they must make judgements about the patient's response to treatment, and interpersonal skills, where they need to judge the patient's well being and reassure patients concerning the progress they are making.

Figure 6.18 sets out the range of activities involved in ward-nursing and shows how they relate to each other. A full HTA would make clear the extent of discretion the nurse has. Nurses must make clinical judgements but the responsibility for acting on these judgements does not always rest with them - as with the neonatal intensive care nurse in Figure 4.7.

The task entails monitoring the wellbeing of patients and judging whether their treatment remains appropriate. It also involves counselling patients and preparing them psychologically for when they return home. Thus, many social skills are involved. It may be possible to model social skills using HTA, using ideas such as monitoring facial expressions and responses, but this is unlikely to be very satisfactory. While this might not be felt to be worthwhile, it is worthwhile setting out what this range of skills might be, what purpose these skills serve and how they must be carried out in the context of other parts of the task. HTA is effective for placing social skills in context just as it places cognitive skills in context.

Management

Management and supervision are important in any organisation since they are concerned with supporting operational tasks. Management is concerned with setting goals, providing and managing resources and ensuring that systems work effectively. Supervision focuses on the issues of ensuring that systems are working effectively, especially that staff are fulfilling their functions. Management and supervisory tasks have not traditionally been dealt with using HTA.

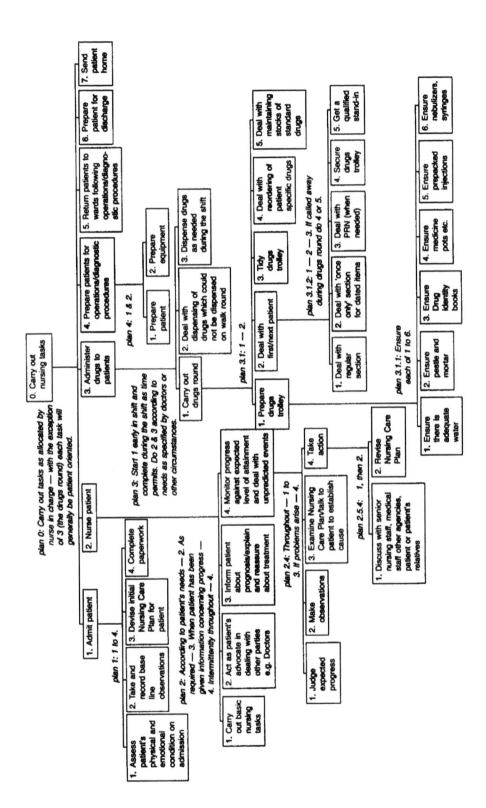

Figure 6.18   Carrying out nursing duties in a hospital ward.

People who are called managers or supervisors in organisations are often given these titles because of historical precedent or the requirement to emphasise status. Often it is arbitrary whether people are called supervisors or managers. We would expect supervisors to report to managers, and managers to senior managers. We would not normally expect managers to report to supervisors, although some managers themselves have supervisors. Management and supervision can only be discussed sensibly by defining their functions.

The management function is to take responsibility for achieving a set of goals in accordance with certain constraints. These goals or targets could be to provide a service to a particular standard or to obtain a profit. Constraints will be concerned with the amount of investment available, a set of resources, including human resources, that are available to use, and various constraints on how things should be done. Constraints include meeting health and safety requirements, complying with local or industry custom and practice and various company standards. Management functions include the following:

- Set up or modify an organisation to meet given targets and goals
- Monitor and maintain the adequacy of that organisation
- Develop the organisation

Depending on the organisation, some of these functions may be required and others not. In a mature company a manager may be required to manage a department by monitoring and maintaining its continued adequacy without looking for changes or without the requirement of having to set up the organisation in the first place. In a new organisation managers may be required to engage more in the processes of developing how things should be done. In a mature organisation under conditions of change a manager may be looking both within the organisation and outside of the organisation for opportunities to improve productivity.

In addition to these central management functions, managers would also be expected to carry out the following:

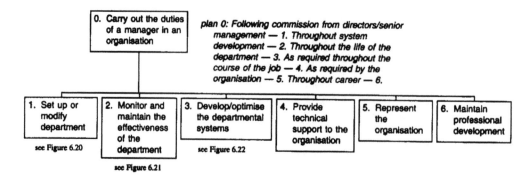

Figure 6.19   The overall functions of the management task.

- Provide technical support to the organisation
- Represent the organisation to outside agencies
- Maintain professional development

These three aspects are not strictly management functions but they are the things that managers often have to do. A manager may have technical expertise that could be used elsewhere in the organisation. Thus managers often join working parties within the organisation to facilitate other developments. Organisations exist within communities and so managers often undertake duties outside. These could be representing organisations at trade associations, industry working parties, representing the organisation to outside inspectors and also supporting public relations. A final function is that managers are also regarded as senior professionals within an organisation and it is their responsibility to maintain their own professional development. One would expect managers to keep abreast with what is happening in their field in order that their own work can be improved.

To represent the management task as an HTA, it is useful first to consider the ways in which the general functions of management can be set out. An example is given in Figure 6.19. This contains the various management functions described earlier together with a plan which sets out the conditions when they are required. Of course, the details of Figure 6.19 will vary between organisations in accordance with how the term 'management' is actually used.

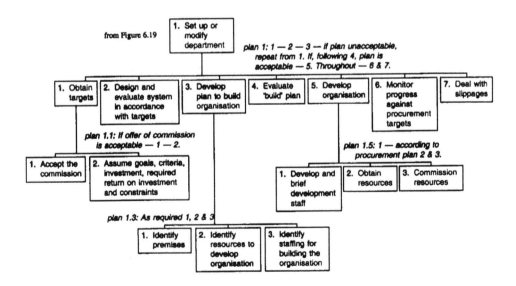

Figure 6.20 Setting up and modifying a department.

Set up or modify department

The subtask of 'setting up or modify a department' is set out in Figure 6.20. When a new organisation, department or project is set up within an organisation the manager charged with this responsibility will first obtain the targets that have to be met or that are expected to be met by the director (or senior manager), who decides on investment priorities. The decision to set up a new project may have been made following some kind of analysis of markets or needs. Therefore, the manager will be given targets to achieve, criteria of performance to be met, will be told of the investment that will be available, and should be told the required return on investment. The manager should also be told, or already know, the constraints. Constraints can include the requirement to use existing plant, the requirement to use existing staff or recruit from the locality, the requirement to use certain raw materials or suppliers and various time constraints for achieving goals and intermediate goals.

A second activity concerning design in evaluating the system in accordance with targets requires the manager and colleagues to create an appropriate system to meet these targets which is consistent with these constraints. This entails analysing the functions of the system, identifying the components that need to be in place to achieve the system and ensuring that these components relate to one another. In effect, this is *systems analysis*. It is something that can be done very well using HTA.

Having generated a prototype system, the manager and colleagues need to develop the plan to build the organisation that will be capable of sustaining this system. This could include identifying premises, identifying resources to develop the organisation and identifying staffing for building the organisation.

The manager and colleagues should now evaluate the plan's likely success. This entails projecting expenditure against the procurement plan; as various stages within the plan are reached, various expenditure must be made. Done properly, this step should enable the manager either to report that the project is or is not feasible, or that it needs modification.

Given the go-ahead, the organisation is developed. This includes developing and briefing staff to assist with the development, obtaining resources according to the development plan, and then commissioning resources as they are obtained. As more resources are obtained then more parts of the system can be tested in conjunction with each other, until the whole system is built. At this stage the infrastructure for normal operations can be put into place and normal operations can commence, thereby obtaining a return on the investment. Throughout the procurement plan, progress should be monitored against targets and slippages dealt with.

Many aspects of this phase of management can be shared between colleagues with different expertise, including financial, technical and human resources staff. Organisations vary substantially with regard to who is employed and who is available for consultation and collaboration. Small companies have different opportunities to large companies. Companies with adequate resources may be able to contract out certain aspects of these processes, whereas less well-off companies cannot. Despite these

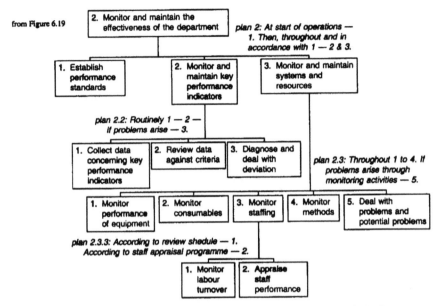

Figure 6.21   Activities in monitoring and maintaining the effectiveness of the department.

variations, the same functions must be met, whatever the size of the project and whatever size the organisation.[31]

Monitoring and maintaining the effectiveness of the department

The manager must ensure that a system continues to perform according to required standards. These standards might be given by directors initiating the project, or they might be given by an outside agency, for example, central government. Performance targets will vary in accordance with the nature of the work involved. In any case, the manager's job is to devise ways in which these standards can be measured and set up systems to collect suitable data.

In addition to targets formally set, the manager should also identify factors that will signal whether or not the department is working effectively. For example, a company may be concerned with output, but it may also be the case that labour turnover or morale in certain departments is a factor which is known to affect output. Collecting appropriate data in this way can help in any subsequent investigation of weakness and help to avert problems that might arise. Thus, it is usually worthwhile to have systems which monitor and maintain key performance indicators concerned with the *product* of companies or departments, and systems which monitor and maintain systems and resources concerned with the *processes* of securing these key performance indicators. Figure 6.21 sets out a range of product and process system measures which may be used to give warning of problems to enable remedial action to be taken. These may be measured and acted on formally or they may be used intuitively by supervisory staff in ensuring the department performs most effectively.

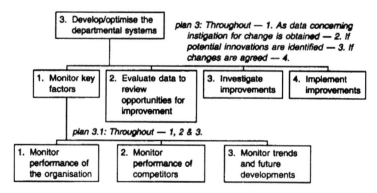

Figure 6.22   Developing the department's systems.

Developing the organisation - managing change

In any system, symptoms which highlight deviation either show that various characteristics need adjustment - this has just been discussed with respect to 'monitoring and maintaining the adequacy of the department' - or they suggest that the system itself needs modification. Where a manager is charged with the responsibility of looking for opportunities to optimise an organisation, he or she must monitor the performance of the organisation, and must also monitor those aspects which could give clues to beneficial change. Monitoring performance of the organisation will show the extent to which the targets expected of a department or system are failing, or the extent to which there is a downward trend in productivity.

As well as comparing how the department is doing it is also important to monitor the performance of competitors, to see how successful they are, to look for new services they are offering and to appreciate the new methods and technologies they are using. It is also important to keep an eye on trends and future developments. In a manufacturing organisation new materials or production methods will substantially affect productivity. In a local authority social services organisation, knowledge of an impending closure of a large factory could prompt an increase in demand for services caused by unemployment. Equally knowledge of developments on the use of the Internet by a sales organisation should prompt an investigation into alternative ways of selling. Activities involved in development of a department are shown in Figure 6.22.

If an organisation is to be responsive to these changes, it is important for a manager to set aside time to review improvement opportunities. Such an activity will signal potential new projects which themselves need to be investigated and, if appropriate, implemented. These decisions would prompt the activities concerned with setting up or modifying the organisation discussed previously, as indicated in plan 0 in Figure 6.20.

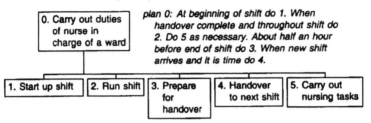

Figure 6.23   HTA of nurse in charge of a ward — an example of staff supervision.

Staff supervision - nurse in charge of a ward

Figures 6.19 to 6.22 represent management *functions*. Whether people who do these things are called managers, supervisors or team leaders, the same functions apply. Some functions are concerned with change, while other functions are concerned with ensuring that a department operates the way it is designed to. Jobs which are mainly concerned with ensuring that a department functions as it is supposed to are often called *supervision*.

To provide an example of supervision, we turn to the role of a nurse in charge of a ward. The nurse-in-charge will be carrying out duties as set out in the HTA in Figure 6.23 to oversee the work of nursing staff whose tasks are set out in Figure 6.18. To supervise staff a supervisor must be clear about what the department is expected to achieve and must be familiar with the capabilities of staffing and other resources.

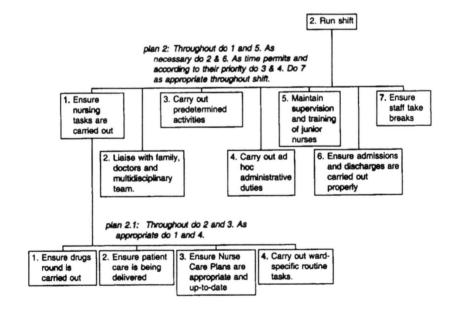

Figure 6.24   Running the shift.

The nurse supervisor's job is carried out over a shift commencing with a set of shift 'start-up' activities which are followed by ensuring that nursing tasks are properly carried out by the shift team. Towards the end of the shift, steps are taken to prepare for handover. While the shift is being run, the supervisor may also need to supplement the effort of staff by undertaking nursing tasks in their own right. This is typical of the demand placed on supervisors - while their main job is to ensure that a shift is run successfully, this can require supervisors to step into the breach themselves. Sometimes a supervisor must contribute in this way to cope with events that have overstretched members of the team, and sometimes the supervisor has specialist skills that no other team member currently possesses. The danger is that a supervisor is too eager to engage in the tasks of team members at the expense of attending to supervisory duties.

Running the shift

Running the shift, set out in Figure 6.24, entails overseeing that staff are conducting their duties appropriately (operations 2.1 & 2.6), ensure that staff have the opportunity to relax (operation 2.7), carrying out various senior tasks (operations 2.2, 2.3, 2.4 & 2.5). One of these is 'maintaining supervision and training of junior nurses'. This is typical of the responsibilities of supervisors. To do this, the supervisor needs instructional skills, one of which is appraising the needs of new staff and the current opportunities available on the ward to meet these needs.

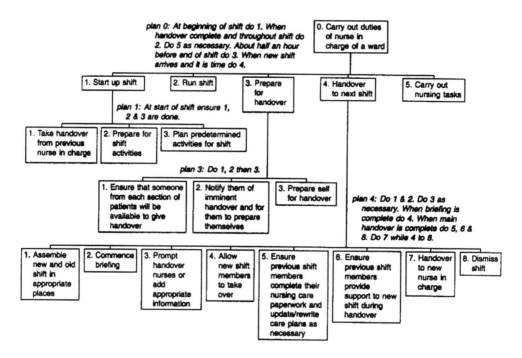

Figure 6.25   Liaising with other shifts.

Liaising with other shifts

Since patients need 24 hour a day nursing care, an essential part of providing nursing care is maintaining continuity between shifts. This means ensuring the team is effectively engaged at the start of the shift and assists the next shift in engaging with the task.

Figure 6.25 illustrates the activities engaged in maintaining continuity of care. In starting up the shift, the nurse in charge must listen carefully to what the previous nurse in charge has to say, and then make plans for the shift to come. As the shift progresses, all staff need to keep notes on progress and significant events; as the shift nears its end, the nurse in charge ensures that the nursing team is preparing itself for briefing the next shift. Handover in a ward is a team effort. The nurse in charge must coordinate this activity but relies on individual nurses with direct experience of past events to brief the new shift. There are also activities to ensure that no loose ends are left as one shift hands over to another, so paper work and other tasks need to have been completed. It may also be necessary in some circumstances to stay with the new shift to ensure that the newcomers are clear about what they are doing. Finally, a formal assignment of the responsibility for the ward must be given to the new nurse in charge such that lines of authority are entirely clear.

Many of these supervisory activities are common in other contexts. They will, of course, vary in detail with regard to local requirements. There is a danger that supervisors are merely seen as 'super' operators, but clearly their job is to ensure that departments work effectively, rather than being engaged in tasks directly. The supervisor is involved in a number of activities, including monitoring persuading, criticising, training and record keeping. These entail a wide range of operational skills, administrative skills and interpersonal skills.

Concluding remarks

This chapter has illustrated the application of HTA to a wide range of tasks, choosing examples from industry, commerce and the caring professions. This has been done to illustrate the breadth of application of HTA. Reviewing these examples, it is clear how tasks from different domains are similar in many respects to one another. A common feature is the extent to which some tasks are concerned with maintaining a system's status. Thus, continuous process control, air-traffic control, nursing, many aspects of management and most supervision is concerned with monitoring a system to determine whether it is performing according to target and whether intervention is necessary. This work paradigm is common in industry and it is also common in management. Managers and supervisors are often concerned with making sure that a system is operating according to intention. Thus, the system must be monitored against targets - if system performance is unsatisfactory then remedial operations must be implemented.

It is important to observe the distinction between *tasks* and *jobs*. Tasks relate to meeting a system's goals, while jobs may entail several tasks that an operator is employed to carry out. To understand jobs it is important to identify all the things that are done when a person gets to work and then leaves work, as well as what goes on in between. This is especially important in jobs concerned with continuity of system supervision or care. Equally, some jobs may only be understood by analysing several different tasks that the job-holder carries out.

Chapters so far have concentrated on how HTA is carried out. In the next few chapters we shall consider how HTA can be applied.

Chapter 7

Making human factors design decisions within HTA

As HTA progresses, the analyst needs continually to consider possible solutions to ensure that the task can be carried out to a satisfactory standard.

In considering design hypotheses, the analyst can first consider the manner in which events and circumstances affect how the task is carried out. These include how events to be dealt with are presented to the operator and the competency the operator must bring to the task. These considerations then provide a clearer insight into how the analyst can suggest ways of managing these factors, for example, whether jobs are reorganised to effect the functions that a job must deal with, how the rate of occurrence of events can be influenced, how the task is represented to the operator via the interface, and how operators are helped in responding to these events through training and support.

Introduction

Task analysis methods serve little purpose unless they help make useful decisions concerned with developing or evaluating aspects of human factors and human resource management. Systematically addressing issues of human factors design is fundamental to HTA. It can be argued that an HTA is complete only when hypotheses have been generated to deal with each part of the task. To provide better support for making design processes, therefore, we need to go further. We shall consider four issues concerned with human factors design and HTA:

- considering the design options
- making design choices
- developing detailed design
- human factors design decisions within the system life-cycle.

Considering the design options

When we look for solutions to human performance problems, there are usually several alternative ways of improving matters. For example, training will enable operators to deal more effectively with a given set of circumstances, but, equally, improving the design of the interface or simplifying the job in some other way, may enable the task to be carried out more easily, thereby requiring less skill. The task analyst may consider a combination of different methods for achieving the improvement in performance. There is a danger that human factors designers restrict themselves to only a few alternatives - often ideas with which they are most comfortable. Thus, one analyst might pursue training solutions to the exclusion of other possibilities, while another focuses on personnel selection solutions, and another may favour interface design. In reality, there is usually a wide range of alternatives and several should be taken together.

Making design choices

When different human factors design alternatives have been considered, the analyst must then select which ones to pursue, then schedule how they will be pursued. The analyst must judge which combination of design features are warranted for the particular task and its context and also be prepared to provide a justification for making this choice. This should form the basis of a recommendation to the client. The final choice is the responsibility of the client, who will bear the cost and manage the consequences.

Developing detailed design

Having agreed a programme of design, details for each step need to be developed. In human factors work, this means specifying such things as how jobs are organised, how information should be provided to operators and how training materials should be developed.

Integrating human factors design decisions within the system life-cycle

For any practical application of human factors, the design activities must integrate with other decisions taken in the development of the system, for example, the design, procurement or installation of equipment. This is important because human factors decisions depend on good task analysis, and task analysis depends on obtaining information that is relevant to the task. This means that different phases of task analysis must be carried out in step with other aspects of system design. If the task analysis and the attendant design is undertaken too early, then the correct information will not be forthcoming. If this design effort is left too late, then further design decisions will be constrained by other non-human factors decisions. If task analysis and human factors design can be integrated with other system decisions, then best use of information and design opportunities can be taken.

Considering the design options

A task analyst concerned with devising design options must first think in terms of the functions to be served by the design; then he or she needs to be imaginative about ways in which design options can serve these functions. Eventually, these design options must be explored to ensure that the design suggestions to be pursued are compatible with one another, consistent with design constraints and acceptable to the client.

In human factors and human resource management the design options are the things that can be changed in order to influence human performance. These things include *interface design, personnel selection and training*. There are also some less obvious things that can be done to support performance, such as improving the reliability of equipment in order to reduce the reliance placed on human performance. These factors interact. We can see this by representing them within the operator-system interaction diagram which was introduced in Chapter 1. Figure 7.1 shows how basic task elements interact to affect skill. These include:

- the schedule of events with which the operator must try to cope
- the availability of inform   control facilities
- the skills the operator possesses
- the working goals the operator is given.

Figure 7.2 shows how the interactions in Figure 7.1 can be managed by typical human factors and human resource interventions. An important practical point is that we are confined to managing these factors that influence experience - we cannot influence the experience factors directly. Thus, Figure 7.2 shows how factors such as team and job design, interface design and training influence performance.

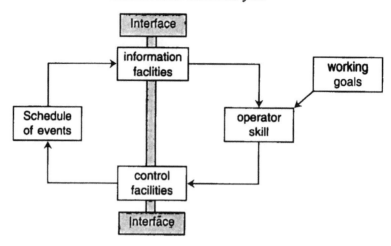

Figure 7.1   The interaction between task factors.

Schedule of events

The *schedule of events* with which the operator is required to cope affects workload, familiarity and diversification. *Workload* issues include circumstances where the operator has more things to do than can be reasonably coped with. One aspect of workload reflects the *rate* at which events and signals are presented. In the automated warehousing task, described in Chapter 4, engineers discussed methods of speeding up the conveyor belt to process orders and paid insufficient regard to the extra demands this placed on the operator trying to make decisions. In these circumstances training would have to be investigated and performance measured to determine how quickly operators could make decisions as this would constrain by how much the speed of the conveyor belts could be increased.

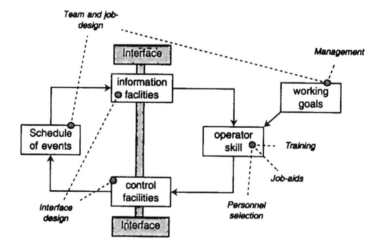

Figure 7.2   Strategies for managing the operator's experience.

While workload issues are sometimes concerned with signals and events occurring too rapidly, events that are encountered infrequently create *unfamiliarity* for the operator. This means that strategies for dealing with unfamiliar events will not be developed and rehearsed.

Operator performance problems associated with scheduling of events which cannot be overcome by better training can sometimes be dealt with by engineering solutions - although engineers may be reluctant to do this as it may incur substantial expense. If task analysis is undertaken sufficiently early in a system's development, then management is more likely to take note of these possibilities, because there has been less expenditure and personal investment on the project. For example, during the construction of a large chemical plant, the task analysis revealed the need for operators routinely every 3 weeks to take a distillation column off-line, in order to clean out one of its associated vessels. The clean-out operation would take half an hour. By applying HTA, the engineers were able to make the clean-out procedures explicit and realised this process would entail substantial shutdown of part of the plant. The shutdown and the subsequent restart was estimated to take 3 days in total. It was recognised that losing 3 days production in this way every 3 weeks was absurd. This focused the engineers' attention onto the problem of avoiding shutdown. This was achieved by inserting 12 metres of pipe to enable product to be recycled while the clean out took place. Many such examples can be encountered through the application of task analysis and their benefits can be substantial.

There are other 'engineering' solutions which do not entail modifying the physical arrangement of elements of a system, but rather depend on changing the *functionality* of the system. For example, some wordprocessing packages are easier to deal with because they have been designed to incorporate fewer features than more advanced versions produced by the same manufacturer. Another example can be seen in motor garages that specialise in fitting tyres and exhausts. In these garages, staff do not need the same range of skills as mechanics in more general purpose garages. By attending to a limited range of products and problems they are able to maintain their skills and expertise more effectively. (Admittedly, the motivation to organise a garage in this way is usually taken to enable bulk purchasing of parts and maintain more comprehensive stock, but there are benefits regarding human factors and human resource management as well.) In a further example, a management team, frustrated by the inability of their design engineers to resolve technical problems, decided that it was better to accept reduced productivity. Lowering performance criteria is always an option for management and one which has direct implications for the pressure placed on operators.

Interface and control

To carry out a task, the operator depends upon information about events and conditions, including task feedback information that can be used to regulate performance. The operator is also dependent upon having access to appropriate controls, suited to the task in hand.

Information requirements

In making any decision, the operator must use information to resolve uncertainty. A person might come to rely on a scar on the chin to distinguish between identical twins. If one of the twins grows a beard, then there is a strong likelihood of mistaking the identity of the twin with the beard until the other twin is also seen. Equally, if a patient presents symptoms of stomach pain to a doctor, the doctor cannot know whether this is indigestion or something more serious without asking further questions and taking more tests. Acting on any judgement with insufficient information is a risk which may or may not be worth taking. An operator cannot reduce this risk simply by considering the information more carefully; if information is insufficient to distinguish between alternatives, then it needs to be supplemented by further information. This may be achieved by training the operator to use additional information. Often, though, sufficient additional information is unavailable because its importance was not recognised when the system was originally built. HTA serves a role here in helping the analyst to identify which information is crucial in carrying out a task. This is indicated through plans and through the feedback that operators need to regulate their actions.

Environmental ergonomics issues

A related issue concerns the adequacy of the environment in enabling signals to be detected and different critical values of signals to be discriminated. For example, if the level of illumination is too low, then different colours and fine detail cannot be detected. Distinguishing fine detail includes distinguishing between different characters on displays. If auditory signals are too weak or are masked by background noise, then crucial sounds may not be distinguished. Thus, this information may be available, but it is not in a form suited for human detection and discrimination. These are problems that will be resolved by basic ergonomics[32]. HTA serves only to indicate which aspects of information display are critical and, therefore, must be given proper attention.

Individual differences in information detection and discrimination

It must also be remembered that people vary in their capability for dealing with different sensory modalities. People who are hard of hearing or who have poor eyesight will be disadvantaged in some tasks. Conditions such as colour-blindness and tinnitus are particularly problematic. Generally, good environmental ergonomics would try to avoid presenting information that compromised known sensory defects. For example, a signal that turns from green to red to indicate danger, without any other means to distinguish these states, is useless for the significant number of people who are red - green colour blind.

Layout of workplace

Although sufficient information and controls to carry out the task may be available, they may be inaccessible. A task analysis may show that two items of information must be considered, but to give attention to one requires the operator to be in a place remote from the other. This means moving backwards and forwards between the two locations. When this happens, the operator may be distracted and forget important information.

Moreover, the status of one item may change as the other is monitored. The same problems can occur in hierarchically organised computer systems. The operator may need to go to different screens to establish the current values of two or more items of information prior to making a decision. In each of these cases, the identification of crucial information through HTA can serve to highlight where these problems might arise.[33]

Representation of information

The manner in which information and controls are represented to the operator has an important influence on the ease with which the information can be interpreted and used. If information and controls can be represented in ways that suggest their use, then the operator will need far less skill than if the interface were poorly designed. These are common ergonomics principles and have been investigated extensively in human - computer interaction research. Many principles for designing interfacing tools - information displays and controls - rely on *functional design*, where the operator is prompted on the identity of a tool and the use that can be made of it by providing a label or an icon which suggests its use. HTA can assist the process of identifying interfacing functions.[34]

Operator skill

Invariably, operators have to learn something of the operations they have to carry out. Sometimes operators or users can learn simply through experience. If interfaces are well designed and the operators know what they wish to achieve, then the correct operations are often prompted by a good interface. Resultant errors should be easy to detect and rectify. Thus, in many computer applications, appropriate actions are prompted by good icons and file-names. Any errors made by the user can be rectified through an 'undo' function which returns the system to its state prior to the erroneous action. In many systems however, engineering the interface in this way is not possible and errors cannot be recovered so easily. So, people must be helped to master skills.

Human resource measures

There are several ways of helping operators bring appropriate skills to the task. One is to *design jobs* in an appropriate manner. In this way, it can be possible to arrange matters such that people can then be recruited who already possess some of the skills to be taught or who have some of the personal attributes that will help them to learn. *Job descriptions* are also helpful because operators then know what is required of them and what is not. They are more likely to engage positively in those parts of the task expected of them and, therefore, experience the various aspects of the task that will help them to learn. Finally, *personnel recruitment* enables staff who already possess some of the required skills or attributes to be recruited. We can look to HTA to make the various aspects of the job sufficiently clear such that jobs can be designed, job descriptions written and personnel requirements for new recruits set out more clearly.

Ensuring competency

The main ways of helping people carry out unfamiliar skills is through an effort to *train* and to provide some form of *job-aiding*. Training may be aimed at ensuring unaided performance or it may be aimed at showing people how to use a job-aid. Training specification is one of the major outcomes from a task analysis project. Poor performance can be the result of a variety of different weaknesses that might be overcome by training. It might be revealed that operators do not properly understand their role in the organisation, or they may not understand the performance targets they are expected to meet. In some situations, operators are required to work with flexibility, but they may have insufficient knowledge and experience to help them deal with the unexpected. The solution could be to teach appropriate knowledge and provide the opportunity for practice in, say, a simulator. The situation might require the operator to recognise patterns and detect changes. The operator may not be sufficiently fluent in carrying out basic actions. And finally, the operator may be unversed in team working and may need to master team skills and become familiar with working with colleagues. HTA can be used to specify training needs and to establish effective training conditions.

In some situations, there is time and opportunity for the operator to access some form of support document (i.e. a job-aid) or help-facility in order to guide certain aspects of performance. Where this is possible and where the requirements of the task are suitable, encouraging operators to use job-aids may be entirely satisfactory. Normally job-aids would be suitable where a stereotyped response to a limited number of situations is required. It is likely to be less appropriate to propose the use of job-aids where operators have to demonstrate flexibility and judgement. Where job-aids are appropriate, it is still the case that the operator will need some training. Such training includes learning how to use the job-aid, gaining confidence in the job-aid and making sure that each action prescribed in the job-aid can be carried out effectively. HTA can be used to guide the analyst in identifying parts of the task where job-aids might be appropriate, showing how job-aids are related to training and helping to design the job-aid itself.

Minimising distraction

The execution of skills is affected by staff being interrupted or distracted. If such interruptions are necessary, then consideration should be given to job design issues. For example, a senior doctor who must concentrate on diagnosis and treatment planning for a patient in crisis, should not also have to manage the detail of other aspects of treatment or in peripheral housekeeping issues. In such circumstance, there is a need for a gatekeeper, for example a senior nurse or senior house officer, to adopt the explicit role of protecting the doctor from interruptions, at least while the clinical crisis persists. Such problems are commonplace and can also stem from the inappropriate layout of buildings, where staff on crucial assignments are interrupted by other people passing through. HTA can be used to examine such jobs with a view to identifying those parts of the job that are distracting and which can be given to others in the team.

Making design choices

So far the discussion has focused on the way in which different aspects of the task system combine to affect performance, and on the different steps that can be taken to ensure adequate performance. The problem facing the analyst is in knowing how best to configure different parts of the system in order to achieve the best results - in terms of effectiveness, efficiency and economy.

Human factors guidance is often presented in the form of checklists, guidelines or design methodologies. *Checklists* guide the designer in looking at different features of an object being designed to establish whether they conform to some ideal. *Guidelines* specify how certain parts of a design are furnished. Checklists and guidelines may help to deal with oversights, but they risk stereotyping the design. The danger with stereotyping is that it can fail to anticipate interactions between elements which may present special problems or opportunities. Design *methodologies* can be different in that they may guide the designer through steps to be considered rather than prescribing how steps should be dealt with. In truth, the process of deciding how elements of an object should be considered in design is often more subtle than checklists or guidelines permit. In human factors design, the interactions between elements of the task can affect each other in unforeseen ways.

Design is a skill which arises from the designer assembling functional requirements and the constraints that need to be observed, then proposing a design that complies with these. The following is an example, which shows how these elements interact in dealing with a design decision encountered through HTA. The purpose of this example is to illustrate the typical richness of real problems encountered. In real situations, tasks can often be accomplished by different methods with design solution involving trading off human factors in different ways.

A practical example of dealing with redesign opportunities

The example refers to the execution of a critical operation in a batch chemical process plant which manufactures synthetic resins. In batch processing, profitability is often determined by how quickly a batch is completed. This leads to more batches per year and, hence, greater profit. In analysing such tasks, therefore, the analyst is alert to those operations whose duration is affected by skilled performance. The batch process under consideration, is that described in Figure 6.3 to 6.5. The vapour emerges from a pipe at the top of the reactor then condenses into the distillate receiver. The process of vaporisation is called 'refluxing'; subgoal 7 is called 'maintain reflux'. This plant was somewhat primitive in terms of process control technology, but it is useful for the present illustration. The plant was arranged on two floors. On one floor the operator prepared batch materials and did most of the control panel work. The control panel was situated next to the reactor, directly above the distillate receiver.

The batch is brought to reflux by operating a steam control on the control panel. Once the reflux has started, the operator's task is to maintain the reflux rate as close as possible to 10 cm every 15 minutes. This rate is controlled by the steam pressure. If the

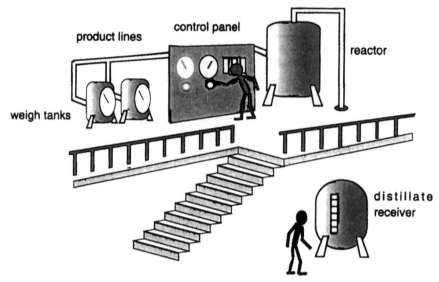

Figure 7.3   Layout of the batch plant.

steam pressure is too high, then the reflux rate will exceed 10 cm per 15 minutes and the batch risks being destroyed through chemical degrading. If the distillation rate falls bellow 10 cm per 15 minutes, then the batch will take longer to complete. Efficient operation is where the distillation rate is kept as close as possible to 10 cm per hour without exceeding it.

The redescription of refluxing is shown in Figure 7.4. To control reflux, the operator is required, every 15 minutes, to go downstairs to establish by how much the level in the distillate receiver has increased. This is done by looking at a sight-glass on the side of the vessel - a sight-glass is a small window that shows directly the level of liquid in the vessel. If the increase is 10 cm, then the operator returns to the floor above to continue other tasks. If the rate has fallen short of 10 cm per 15 minutes, then the operator is required to return to the control panel, calculate the change required to increase the rate, make this change, then monitor the consequence after 15 minutes. If

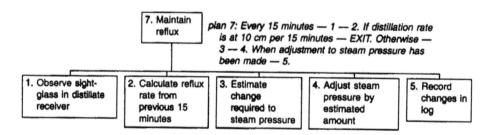

Figure 7.4   Redescription of 'reflux'.

the rate is above 10 cm per 15 minutes, then the operator must proceed to the control panel with greater haste to reduce the steam pressure by an appropriate amount to try to avoid ruining the batch.

The operator's skill, in this respect, rests with the conscientiousness with which the distillation rate is monitored and the accuracy in calculating the adjustment to the steam pressure. Management wanted batch times to be reduced in order to improve productivity. They judged that improvements to the distillation rate would enable this. The analysis shown in Figure 7.4 is an ideal representation of how the job should be done. It was clear in practice, however, that operators consistently fell short of this ideal and failed to monitor the sight-glass as frequently as they were supposed to. As they could not afford to risk batch quality, they deliberately fell short of the optimal · distillation rate to avoid the risk of overheating, hence batch processing times were longer than they should have been.

Management assumed that because this problem was associated with operator skill, then its solution must lie with training. There were two main elements of the task which affected optimisation. First, the frequency of monitoring the distillation rate was critical. If for some reason operators did not monitor frequently, then there was a risk of failing to detect an excessive distillation rate that could ruin the batch - hence their suboptimal strategy. This prompted the analyst to consider what could cause this reduction in the frequency of monitoring. The obvious reason was that operators were expected by management to carry out other tasks whilst carrying out distillation - the HTA in Figure 6.4 specifies that the operator also needed to attend to making preparations for the next batch and to maintain the progress of other batches. This work was excessive and did not permit sufficient time for conscientious monitoring of the sight-glass, especially since, to monitor the sigh-glass, the operator had to go to a part of the plant which was never visited when attending to other activities. One immediate design hypothesis considered was that a level indicator displaying the level in the distillate receiver should be installed on the control panel, closer to where the operator was engaged in other activity.

The second main element of this task which affected optimisation was concerned with how accurately the operator could predict adjustment to steam pressure from the observed deviation of distillation rate. The operator had to work out how much steam adjustment was required, taking into account the previous steam pressure. It was assumed that this skill could be improved by some form of simulation training where the operator practised this control relationship. Unfortunately, the reality was not so simple. In order to maximise performance, management only required a thorough clean-out between batches when the chemistry of two successive batches differed significantly. Normally, the products manufactured were consistent with one another, so the clean-out amounted to a quick 'flush-out' with hot water without any extensive purging procedure. This meant that as successive batches were processed, there was a buildup of product on the inner surfaces of the vessel and on the steam coils. Hence, the control relationship changed gradually from batch to batch. It was possible that an operator could master the basic control relationship and then make adjustments when coming on shift. However, a further consideration was that batch times varied between 15 and 24 hours and the shift cycle was 12 hours. This meant that when an operator came on shift there was no guarantee that the distillation operation would be encountered. A further complication was that there were four such batch plants and operators rotated between them. Indeed, it could be a long time before an operator again encountered the distillation operation on the same plant. The conditions for exploiting and updating a control relationship in these circumstances were very unfavourable. The measures that would need to be taken to stabilise conditions such that learning a control relationship would be worthwhile were not acceptable by management.

The conclusion from the analysis was that training could not provide a solution to this problem. Nor was representing the distillation rate in a more sensible place likely to be much help, given that these skills could not be acquired to a sufficient standard. The only sensible option for improving batch performance was to install *automatic control*. This would cause steam pressure to be adjusted continually in accordance with changes in the distillation rate. This involved some cost and inconvenience, including interruption to the production schedule. The management were not prepared to countenance such inconvenience at this time. The outcome of the exercise was that management did nothing. They had to accept the responsibility for suboptimal batch times and that this was the price that had to be paid to avoid batches being ruined by excessive heating.

Despite its unsatisfactory conclusion, this example shows a number of useful things concerning how analysts need to think broadly in considering hypotheses. The unsatisfactory *schedule of events* was crucial in showing how any learned skill was unlikely to be retained. Shift arrangements, batch cycle times and allocation of staff to different plants, reduced the frequency of encountering the operation and, hence, reduced opportunity to rehearse a consistent skill. It was also interesting to note that, critical to the operation in focus, was another distant operation, namely 'clean-out', which directly affected the schedule of events. Had thorough clean-outs been endorsed, then the characteristics of vessels would have remained consistent and control actions could have been administered with more confidence. Indeed, to provide the conditions under which the skill of steam adjustment during distillation could be retained, the management would have endorsed a policy where thorough clean out was always undertaken, where operators remain within the same plant and where batches always started at the same time as shifts, such that operators would consistently encounter the same demands. Of course, many of these demands would not make sense in a commercial environment. On the other hand, it is also clear that management must accept the consequences of their decisions. It is tempting for managers to maintain that performance problems can be solved by training or personnel section rather than engineering modifications, simply because this is most expedient and less intrusive. The example also demonstrates how the analyst must open up the options which affect performance in order to entertain a wide range of alternatives. Then the options must be closed down, first by observing the real constraints on the task, and second, by accepting the ruling made by the client - albeit, following suitable negotiation.

Context and constraint and design decisions

The example of the refluxing operation just discussed emphasises the importance of appreciating task context in understanding how skills are affected and how design hypotheses are constrained. We have discussed several of these, but there are a number of other issues to consider. A fuller list is as follows:

- Difficulty of the task
- Predictability of events
- Controllability of events
- Frequency of events
- Severity of consequences of error
- Representation and feedback
- Recoverability
- Stressors
- Access to help
- Environmental and movement constraints
- Costs of training support
- Legal, industrial and cultural compliance.[35]

The factors will now be described separately, but it is important to emphasise that their effects on design choices can only be considered by assessing all relevant factors in combination. It is only when we see them operating in combination can we properly understand how they affect what the operator or trainee experiences.

Difficulty of the task

The more difficult a task the more time a trainee will need to acquire skills and the more likely he or she is to make mistakes.

Predictability of events

If the set of events which trigger a particular response are predictable then the operator can anticipate and prepare for events. Someone undergoing training can be prepared for learning and freed from other activities, as can an instructor. If the onset of relevant conditions is unpredictable, then no such preparation is possible. Operator response and instruction will be less prepared.

Controllability of events

In training some tasks, instructors may be permitted to exercise control over when things happen. For example, cleaning or purging operations, are often required at sometime during a shift, but they are rarely required at a precise moment. This means that an instructor would be able to schedule the task when it is most convenient to do so. In other tasks, the instructor may have even more discretion. For example, the instructor may be able to safely move a process parameter slightly out of tolerance to enable the trainee to practice a routine to bring the parameter back into line. Such practices must be avoided where there is risk, but where such actions merely provide a temporary reduction in optimal system operation, the cost may be entirely justified in training terms. Many contexts provide opportunities to influence what people deal with at any point in time. This can enable novices to attend only to things they have already mastered and it can help them acquire new skills at a rate consistent with their progress.

Severity of consequences of error

Mistakes at some tasks are trivial or, at least, not life threatening. In such cases mistakes by new operators may not matter and so, people can be taught by on-job training and even left to gain experience on their own.

Recoverability

Some tasks permit recovery from error and some tasks do not. If there is no easy recovery from error and the consequences are severe, then trainees cannot be permitted to carry out these tasks if there is risk that they may not cope - even if there is an instructor available to help them. Fault compensation and rectification tasks fall into this category.

Representation and feedback

The manner in which information or resources such as knobs, dials, escape hatches etc. are represented can significantly affect the skill with which such resources are used. According to their design, such resources may serve to remind the operator what things could be done or what should be done next. The extent to which an object in a control environment prompts the operator in its use has been called 'affordance' by various authors. Similarly, whenever an action is carried out the operator will be looking for feedback to confirm that the action was correct or regulate the action in some way. In some systems, feedback which can be used to regulate performance in this fashion is readily available, possibly deliberately designed into the system. In other systems, such feedback is unavailable, or delayed such that little use can be made of it. Tasks with good affordance and feedback will be easier to master since the novice operator is prompted what to do and receives confirmation that the action was appropriate. Learning may take place on the job, even unaccompanied provided the consequences of error are mild. Where affordance and feedback are unsatisfactory learning in the operational context is unlikely.[36]

Stressors

Some real situations place stress on the operator and this may inhibit learning. This prompts a justification for adopting simulation training, but this must be done with care because it is often the case that people may learn on a simulator but still be subject to stress when working in the real situation. If the task entails a prescribed procedure, then the operator is best equipped with a job-aid and then trained to use the job-aid. However, if the task requires the operator to make judgements concerning unplanned events, then even a decision-aid may not suffice and some form of skills training will be necessary.

Access to help

Any factors which limit access to help in an operational context should be taken into consideration in determining whether operators can be given effective access to support or helped to become self-sufficient. Working in hostile environments often creates the conditions where access to help is limited. For example, a maintenance technician may have to work alone, and cannot ask for advice.

Environmental and movement constraints

Environments in which there is insufficient light to see, or too much noise to hear, or too little space to move with freedom place constraints on the operator's decision-making or choice of action and this will affect the human factors decisions made.

Costs of training support

Cost is a serious consideration in all organisations, and can influence the choice of information and control features, and may also limit the training and support that can be provided.

Legal, industrial and cultural compliance

Often, things have to be done one way rather than another, because there is a law which says they must. Sometimes it is necessary to demonstrate compliance with company standards or industry codes of practice.

Combining conditions

The examination of reflux in the batch plant, described earlier, illustrated how different contextual conditions and constraints interacted to support or challenge different design hypotheses. Table 7.1 takes the list of performance constraints discussed above and shows, for a set of safety critical tasks, how their influence might be combined in reaching design decisions. The different values in the matrix are merely estimated and not quantified, though they might be if suitable data could be collected or estimated. In the absence of such formal data collection, informal estimations are still valuable because they can be used to guide the analyst to consider different combinations of possibilities.

Table 7.2 shows how these considerations could be incorporated into a task analysis table. Each operation of concern could be considered in terms of how it relates to the various contextual factors. On the basis of this, the analyst could draw out appropriate design hypotheses to be entered in the 'comment' column. In principle the analyst could make these estimates routinely as a basis for deciding what recommendations to make. In practice, analysts are more likely to rely on their experience to make these decisions.

Table 7.1   Factors affecting performance and some consequences for their combination.

Task Type	Task difficulty	Predictability of events	Controllability of task	Frequency of events	Severity of error	Task feedback	Recoverability	Stressors	Access to help	Environmental constraint	System response times	Costs of training support	Legal (etc) compliance	Comments
Normal operation to minimise current risk	var		lo		hi		lo				slow		hi	Requires conscientious operation according to appropriate safety principles. Interlocks, procedural guides, screen-based help will assist compliance. Knowledge of consequences of erroneous operation must be understood. Simulation training will enable consolidation of procedures.
Escape and evacuation		lo	lo	lo	hi	lo	lo	hi	lo	hi		hi	hi	These tasks are carried out infrequently under stressful conditions. Frequent realistic drills are required. Computer-based or classroom teaching may alert to dangers. Procedural guides serve as reminders prior to exercises. Regular inspection of emergency equipment and validation of procedures is required.
Supervision and co-ordination of evacuation	hi	lo	lo	lo	hi		lo	hi	lo			hi	hi	Entails knowing where people are, judging suitability of escape routes and, hence, contingency planning. There is a need for good records and display systems to advise on locations of staff. Need for simulated practice to learn to deal with contingencies, then build contingency planning into full drills.
Co-ordination of emergency/relief services	hi	lo	lo	lo	hi		lo	hi	lo			hi	hi	Similar to above, but requires knowledge of plant states and problems of accessing plant areas, including planning mobilisation of emergency equipment.
Using emergency equipment		lo	lo	lo									hi	Detailed procedures training plus opportunity to practice within routine drills. Procedural guides useful as pre-instructional briefing, but evidence must be sought that unaided skill is present.
Supervising plant during emergencies	hi		lo		hi		lo	hi			fast	hi	hi	This entails control and protection of systems. It entails having a good understanding of systems to enable flexible operation, given that availability of plant is unpredictable. Analytical skills needed. Intelligent planning aids may help.
Administering permit to work systems					hi		lo		lo				hi	Entails adherence to company procedures. Diligence, rather than skill is needed. Off-job training can introduce procedures, but these should then be practised in the real context.
Maintaining safety systems	var				var	poor	var	var					hi	Entails routine checks/tests on system availability. Checklists should ensure thorough coverage and procedural guides to ensure adequate checking of items. Personnel undertaking inspection must be qualified and diligent. Thorough records must be kept of work done and work remaining to be done.
Maintenance tasks	var	lo	lo	lo	var			var		hi			hi	Basic craft training plus opportunity to learn/practice specialist skills, often at point of need.

Key: var — variable;   hi — high;   lo — low.

Table 7.2   An example of an HTA table recording the factors affecting performance as an aid to developing human factors hypotheses.

Task Analysis	step?	Task difficulty	Predictability of events	Control of task	Frequency of events	Severity of error	Information representation	Task feedback	Revoverability	Stressors	Access to help	Environmental constraint	Costs of training support	Legal (etc) compliance	Comment
6 Deal with emergencies plan 6: Do 1, then as appropriate — 2 or 3. 1 Assess situation to establish the extent of the emergency	no	hi	lo	lo	lo	hi		lo	hi			hi	hi		*This entails control and protection of systems. It entails having a good understanding of systems to enable flexible operation, given that availability of plant is unpredicatable. Analytical skills needed. Intelligent planing aids may help.*
2 Deal with local isolation	yes														
3 Deal with emergency evacuation	yes														
6.2. Deal with local isolation plan 6.2: 1 — 2. 1 Assess extent of problem	no	hi	lo	lo	lo	hi		lo	hi			hi	hi		
2 Isolate affected area	yes														
6.3. Deal with emergency evacuation plan 6.3: 1 to 5 in order. 1 Press emergency shutdown button	no														
2 Sound alarm	no														
3 Obtain and don breathing apparatus	yes	hi	lo	lo	lo	hi		lo	hi			hi	hi		
4 Ensure personnel are evacuated from area	no	hi	lo	lo	lo	hi		lo	hi			hi	hi		*Entails knowing where people are, judging suitability of escape routes and, hence, contingency planning. Need for simulated practice to learn to deal with contingencies, then build contingency planning into full drills. Scope for screen-based help.*
5 Move to muster area	no	hi	lo	lo	lo	hi		lo	hi			hi	hi		

Collating and resolving design decisions

As an HTA is conducted, so the analyst will assemble various hypotheses about the task and also various constraints about which hypotheses are suitable. At the end of the analysis, therefore, there is a need to collate the ideas collected in order to decide which ideas should be taken further. This will inevitably entail a number of trade-offs to ensure that ideas are compatible with each other. There may also be a substantial amount of negotiation as ideas are rejected. Sometimes ideas are rejected simply because the client is not comfortable with them - these may have been rejected for reasons associated with investment of time, money and reputation. If there are no satisfactory alternatives to the ideas originally presented, then the analyst must look to ways to persuade the client - hence consultancy skills and sensitivity to the client's circumstances are critical.

Cost issues

Cost issues are fundamental and always need to be respected in an applied project. This does not mean that criteria such as safety can be disregarded. Thus the client and the analyst should seek solutions that are economical whilst maintaining standards of safety and productivity.

While costs are considered as the task analysis progresses, a full assessment of cost implications can only properly be considered towards the end of the analysis. Rejecting hypotheses on cost grounds while the task analysis is being conducted is premature because several operations may eventually benefit from the same solution. For example, the analyst might recognise that a number of operations could benefit from a computer-based training programme. Often, off-job training provided by a simulator or computer-based training program is helpful because it provides opportunities for the trainees to practice dealing with different events, including those critical events that are unlikely to occur during a period of practice on the job. But purchasing and programming a personal computer may seem prohibitively expensive, especially when its benefits appear marginal. If, however, at the end of the analysis, several such opportunities are recognised, the investment in the computer-based training solution may be justified.

Other examples relate to the balance between training and workload. Early in the analysis it might be felt that certain complex tasks are best carried out by specialist personnel and so a team redesign solution is favoured, with existing specialists used to deal with the most challenging situations. However, when the balance of task elements is properly appreciated, it might become apparent that it is better to require all staff to deal with unforeseen events because they may be in the best position to appreciate the initial conditions. Thus, nurses or office staff or plant operators might all receive training in dealing with the unexpected, for the simple reason that they are available and have direct experience of what has occurred in the recent past.

Developing detailed design within the system life-cycle

The book has emphasised the requirement for analysts to be broad in their approach and use analysis to explore issues on their merits as they emerge from the task and the task context. Table 7.3 lists the different sorts of human factors considerations identified that can emerge through the task analysis. It is also clear, in real projects, that practical solutions can entail the analyst proposing several of these in order to provide a solution to a client that meets operational requirements and is otherwise acceptable to the client, for example, remaining within budget. Each element of design needs to be managed properly in order to ensure that it is accomplished in a satisfactory way. Specific details of these different aspects of design, as they relate to HTA, will be treated in later chapters. For the present, however, it is important to recognised that there are time constraints concerning when different issues can be dealt with both in terms of the order in which they are addressed and the opportunities within the design of a system when these matters can be addressed.

Table 7.3 is a matrix where the rows are defined in terms of aspects of human factors and human resource design issues and the columns represent stages in the design and development cycle of a system. The design cycle used here is a simple one and could be elaborated in a number of ways. For example, a more elaborate treatment could extend the number of stages of design and it would include operating and decommissioning as additional stages[37]. The cells in the table show, approximately, when the different design issues should be resolved within the design cycle. We see, for example, that *engineering and technological* issues often need to be resolved early on because these are fundamental to the system meeting its design intent. *Environmental constraints* should be investigated early on because this can affect issues such as how many people must be employed and where and how they will work. This will provide input to *job-* and *team-design*, which must precede *workplace* and *interface* design because it is only when an understanding of how people will be used in the system, that their working arrangements can be specified. Once jobs and the workplace have been specified, then decisions can be taken concerning how operators will be supported in their skills, through providing *job-aids* and *training*. When job-aids and training are specified, it will be better understood what sorts of people will need to be recruited, so that the various *human resource* management issues can be put into place.

Even though some natural ordering of these different human factors and resource issues can be justified, it is still necessary to adopt an iterative approach to human factors design, since views on how different aspects of design will be resolved must usually precede any detailed design; moreover as later design decisions are made, it is often necessary to review and revise earlier decisions to ensure consistency or overcome problems that emerged later on. For example, making an early decision to rely on training will influence other design decisions because it may have been assumed that a certain level of operator skills was possible to achieve. However, it may prove impossible to train people to the standards required and so the basic task must be simplified by reviewing earlier engineering, job-design and interface design issues.

Figure 7.5 shows how HTA may be used within this sort of iterative cycle. HTA can be carried out early on in a system design. Insights may be used to influence some aspects of engineering design but, more likely, the engineering design issues identified will constrain the HTA which can then be used to understand aspects of the environment that will affect the task. This will help specify how teams and jobs may be constrained which will influence how interfaced and other workplace issues are affected. Resolving these issues leads to further modification of the HTA which can be used to resolve the development of job-aids and training. When these are specified, human resource issues can be resolved. As decisions are made, the task analysis can be clarified. Also, as problems emerge, it can become clear that earlier design decisions need revision.

Table 7.3   *Types of solution to support task performance and their relation to the system design cycle.*

Human Factors/Human Resource Issues	Development of system concept	Detailed design	Construction of plant	Construction of control facilities	Commissioning
Engineering options					
• Improving reliability of parts of the system under control.			•	•	•
• Automating parts of the system under control.	•	•	•	•	•
• Changing the capacity of parts of the system under control.			•		•
• Limiting the functionality of the system under control.	•	•	•	•	
Workspace issues					
• Identifying constraints on the design of the workspace.		•	•	•	
• Providing a workspace.				•	•
Team and job design options					
• Changing the size of the team.				•	
• Reallocating specialisms within the team.				•	
Interface issues					
• Providing adequate information to support performance.				•	
• Ensuring that information is located in easily accessible places.				•	
• Ensuring that information can be detected and discriminated.				•	
• Ensuring that the interface is represented in a suitable manner.				•	
Training					
• Ensuring that goals and criteria are understood.					•
• Helping operators to plan and make decisions in unfamiliar situations.					•
• Ensuring that operators are sufficiently familiar with different circumstances.					•
• Ensuring that operators can recognise patterns.					•
• Ensuring that operators can carry out actions.					•
• Ensuring that operators have learned to work effectively with one another.					•
Job-aids					
• Prompting operators concerning standards to maintain.					•
• Prompting operators what to do next.					•
Environmental issues					
• Identifying environmental constraints		•	•	•	
• Minimising distractions.				•	
• Ensuring that information is not obscured by a poor environment.				•	
• Ensuring that the thermal environment is satisfactory.				•	
Human resource issues					
• Optimising recruitment to match task requirements.					•
• Specifying job descriptions to clarify roles.					•

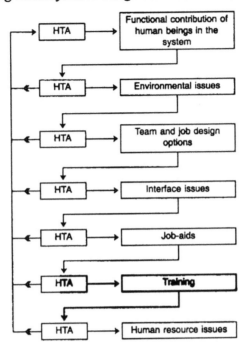

Figure 7.5   HTA within an iterative human factors/human resource design cycle.

All man-made systems have a design cycle, whether they entail industrial engineering or are concerned with a predominantly social organisation. In devising a system for delivering and regulating a social work programme, for example, it is still necessary to understand what the system aims to achieve and then to anticipate the role of equipment and people and how these interact.

Concluding remarks

This chapter has raised the question of how the task analyst makes design recommendations as the HTA progresses. A number of human factors methodologies have been proposed to aid the designer. However, it is stressed here that different circumstances often combine to affect performance in different ways, warranting different sets of design solutions to be proffered. Rather than adopting a procedural solution to human factors design, it is suggested that the analyst should think about the task system and consider how factors interact for an operation currently under consideration. Then these insights can be used to consider how practical steps can be taken to influence and manage these processes. The principal methods that are adopted, together with how they will affect performance, are summarised in Table 7.3.

The problem facing the analyst is that different circumstances influence the task context in different ways. As a consequence, the analyst must think carefully about the aggregate effect of these different circumstances in order to arrive at a set of design solutions most suited to the task and its environment.

Design must always take account of the *cost* of potential innovations. Costs of different hypotheses can be considered as the HTA progresses, but often a proper cost appraisal must be left till the end because costs may be justified when a wide range of solutions are considered together.

This Chapter has focused on the issue of making design choices. This is an integral part of HTA and made complex by the need to consider the interaction between different solutions and the constraints that need to be observed when making these choices. When choices have been made, the analyst must then turn to working out how these solutions may be implemented. This will be the subject of Chapters 8 to 14.

Chapter 8 will consider issues of team and job design.

Chapter 9 will consider issues associated with work design, especially the design of interfaces and providing adequate information to support the task.

Chapter 10 deals with issues of training design and development.

Chapter 11 deals with the design of job-aids and other forms of support documentation.

Chapter 12 demonstrates the application of HTA to supporting various human resources management tools.

Chapter 8

Teams and jobs

Teams are required when people must collaborate to meet a required goal. This arises when task demands exceed that with which a single operator can cope, either because workload may become excessive or where complementary skills are required.

HTA can contribute to understanding team tasks in two different ways. It can be used to develop a functional analysis of the overall team's task, which can then be used to help understand how team members must collaborate. HTA can also be used to help understand the ways in which people relate to one another when they work in teams.

Introduction[38]

One of the concerns of human factors is to evaluate and support teams that already exist and also to devise how new teams could be structured to fulfil functions within organisations. Teams are justified when task demands exceed the capabilities of individuals.

One reason for requiring teams is that there is too much for one person to do in terms of the effort he or she can exert and sustain. This can arise because separate events place too great a demand upon the individual. For example, a single person may be unable to lift a heavy object without help. Equally, an emergency situation may place too many demands on the single operator and, therefore, require that several people collaborate. Teams are also required to deal with those contingencies where too many demands arise at once for the individual to cope, even though the separate demands may be within the competency of the single operator. This means that there must be a sufficient number of people, with appropriate skills, available to deal with the workload that might arise.

A second justification for teams is where the skills required do not reside within a single individual. People with different expertise need to be engaged and must learn to complement one another in dealing with problems. This invariably means each partner recognising when his or her expertise is more suitable than that of colleagues. It also means ensuring that colleagues are properly briefed when their turn comes to take control.

This chapter will discuss how HTA may be used to deal with team design issues. HTA can contribute in three main ways. First HTA is a *functional* method for examining how tasks are undertaken by a group of people. Even within the context of a team, people are working to pursue individual goals, but in a team they must also collaborate to meet common goals. This perspective can be used to understand what is expected of the team and then used to understand how actual team members can collaborate to meet the team's goals.

Second, HTA can be used to provide elements of the task that can be compared in an assessment of workload. Workload assessment is important in evaluating whether a particular organisation of duties within a team can be accomplished, given a certain set of events.

Third, HTA can help the analyst to understand the nature of team skills. In order to operate effectively as a team, staff need to do more that share a common goal. They must cooperate effectively to ensure that their combined effort is effective. There are a number of ways in which people may cooperate with colleagues. These can be thought of as 'team tasks' which can then be examined using HTA.

Collaborating on a common goal

When people work as a team they collaborate to achieve a common goal. This sort of collaboration can be illustrated by reference to several of the contexts already discussed in previous chapters.

The packaging line operation

Chapter 2 included a discussion of a packaging line operator. This example showed how line stoppages were becoming a concern and that management felt there was a need to improve the diagnostic skills of maintenance staff. Understanding the respective training needs of staff was only achieved when the line operator and the maintenance technician were treated as a team. This was accomplished by first considering their joint task. The area where their tasks intersected was concerned with 'deal with system trips and major blockages' as shown in Figure 8.1. This assumes a single entity carrying out the task. In reality, two people were engaged because specialist maintenance skills were required to deal with problems. By examining this task it became clear that operation 3.2, concerned with the maintenance activity is the province of the maintenance technician. Figure 8.2 shows how this task is revised to delegate carrying out maintenance to the technician and Figure 8.3 shows how the technician's task is set out. Thus, the line operator now deals with maintenance by referring the problem, but also has to carry out additional activities and planning in order to convey the necessary information

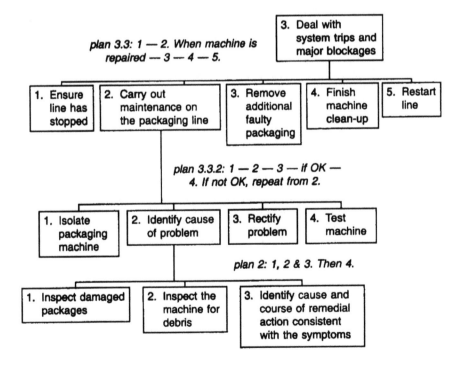

Figure 8.1   A combined analysis of operating and maintaining the packaging line.

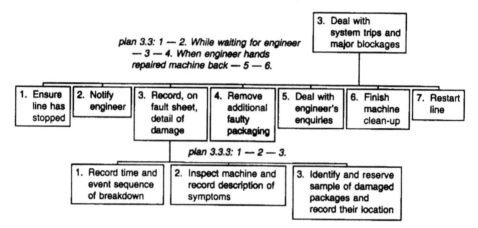

Figure 8.2   The resultant task for the line operator.

to the technician. The line operator's task is expanded as a consequence to ensure that symptoms are preserved and information recorded and conveyed. The technician's task must rely on information being provided and symptoms being preserved by the line operator.

Supervising a railway system

Figure 4.6 described the analysis of an underground railway control task. This HTA was used to explore possible ways of organising personnel in a central control room. The task analysis in Figure 4.6 showed the functions that had to be fulfilled by the control team. One of the implications of running a railway is that traffic develops as the

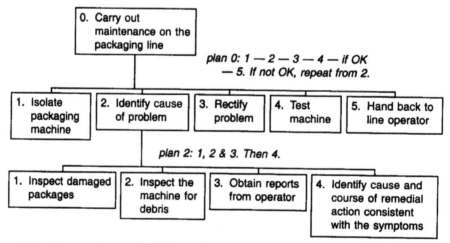

Figure 8.3   The resultant task for the maintenance technician.

day progresses and declines again towards night-time. Over night is the occasion for engineering to be done. Engineering is scheduled to be completed before normal services, with sections of the railway line being returned for normal service as the timetable requires. This daily pattern means that the same workforce cannot be justified throughout the day and night. Also, because this railway is controlled automatically, under normal conditions there was little need for staff to become actively engaged in making control decisions. The job of the controllers was to monitor and to deal with contingencies as they arose. Given this vastly fluctuating pattern of events, how is a team best organised to meet the requirements of operating this railway?

One solution would be to look at team functions, as expressed in the HTA, and delegate different parts to different people. Thus, one class of operator could be required to look after 'provide service' another would 'maintain best service' another would 'deal with system problems', while another class of operator would 'deal with information'. This would not make much sense because the skills required to carry out these different operations interact. For example, as an operator makes another service available in accordance with the timetable, then he or she is better acquainted with the conditions against which to monitor. If the operator was aware of a timetabling problem then this could be taken into account more easily when monitoring best service. If a system problem, such as a train breakdown, arose which required some analysis of the status of the railway in order to work out a recovery plan, then operators previously engaged with the service could deal with the unscheduled event more effectively.

If different team members were allocated these separate functions, four types of staff member, plus their supervisor would need to be continually in communication with one another. Staff would continually demand from colleagues information about the previous status of the system. An alternate suggestion is to create a team of generalists, where each person is engaged on all parts of the railway. Thus each person is trained to deal with the full range of tasks as set out in Figure 4.6. When on shift, they would monitor the system, then deal with events as they arose, talking to one another as the consequences of their actions affected the decisions and activities of colleagues. This would not work very effectively unless there was someone to arbitrate, because staff might start attending to the same events and overlook other events. This suggests an operational role for a supervisor.

Figure 8.4 sets out a specification for the supervisor's job. This job design includes some administrative tasks, such as 'preparing duty rosters' and making sure that sufficient staff are available for duty. The coordinating function necessary to ensure team members match the current demand of the railway is provided by operations 3, 4 and 5.

Operation 3 requires that the supervisor continually monitors the activities on the railway - the supervisor should be aware, at a general level, of all significant events that the control team needs to be dealing with. Operation 4 entails the operator monitoring the team's response - as events arise are team members detecting their occurrence and are they dealing with them effectively? If there is a problem with this response, the supervisor intervenes to direct team members towards these events. Operation 5 is concerned with judging whether there are adequate members of staff currently on watch to deal with the current demand. Thus, as the day progresses and the timetable dictates,

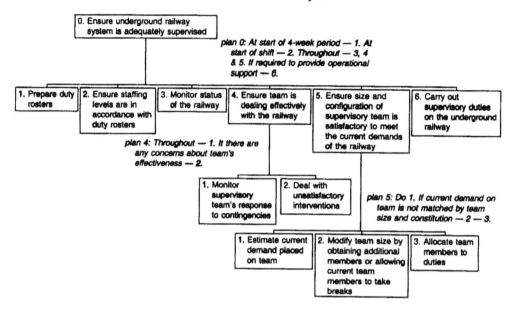

Figure 8.4   A proposed supervisory function for an automatic railway.

the size of the team will be increased by bringing staff into the control room or releasing them. Also, if unscheduled events arise which prompt extra demand, then staff who are currently resting will be brought into the fray to support the team. In any case, operation 6 ensures that the supervisor is also available to fit in with operational support, should the system demand.

Collaboration between nurses and doctors

Figure 4.7 showed the main ways in which teams of doctors and nurses collaborate to deliver effective neonatal intensive care. Of the operations listed, some are clearly the province of nurses while others are the tasks of doctors. For example 'settle baby and make initial measurements' is a routine task, carried out by nurses as babies enter the unit. 'Specify treatment/care plan' and 'revise treatment/care plan' are things done by doctors. Despite these apparent assignments of task to team members, the reality of the situation is that collaboration is called for throughout. A doctor revising the treatment plan relies on information obtained by nurses during their nursing activities. 'Reviewing/ evaluating whether the baby is progressing satisfactorily in accordance with the current treatment/care plan' is the responsibility of doctors, but they must rely on nurses to monitor the ongoing situation and make sensible judgements about the baby's progress before referring problems to the doctors. Indeed, the full analysis of this task shows how virtually all of these operations depends upon information being provided by colleagues - doctors and nurses - and each operation provides opportunities to obtain further information about the baby to communicate to colleagues.

If staff members adopt a strict demarcation and focus on specific goals without appreciating the requirements of colleagues, then they simply cannot carry out their team function effectively. This is emphasised particularly by reference to 'prepare schedule of treatment/care activity for each baby'. This operation is undertaken by a nurse in charge of the ward. This is critical because one of the most valuable acts of care and treatment that a premature baby can receive is to be allowed to rest as much as possible, without unnecessary disturbance. If team members simply pursue individual goals without reference to other activities, then babies will be disturbed continually as different tests and treatments need to be carried out. While these interventions must occur, the senior nurse is able to control when they are done, such that disturbance is minimised.

System coordination also includes ensuring that the system is in a suitable state for colleagues to fulfil their responsibilities. Nurses know that doctors have to administer treatments. If they also have some understanding of the criticality or immediacy of treatment, they will more ably collaborate. If they know something must be done within a certain time frame, they can coordinate their own activity to enable it. Thus, if the baby is to be undressed for a wash or to adjust a dressing, it makes sense to coordinate with the senior house officer responsible for making an invasive test. The same acts of collaboration should be apparent in all systems. All systems need maintenance. If a maintenance act is required within a certain time-frame, the system supervisor may use discretion to identify a suitable time, when system demand is low, for example. Thus, the needs of maintenance will be served and the disturbance to the system is minimised.

Collaboration - some observations

HTA helps understanding the tasks of teams by addressing the joint functions of team members. Each of the three cases cited required the analyst to think in terms of a joint enterprise, rather than considering the tasks of team members separately. In the case of the packaging line operator's task, the required collaboration was achieved by one operator deciding which goal needed to be attained, then assigning responsibility for carrying it out to a colleague. This is a form of delegation, although no status relationship is implied here, as is the case with many acts of delegation. The principle is a simple application of HTA in team and job design and is illustrated in Figure 8.5. The left-hand

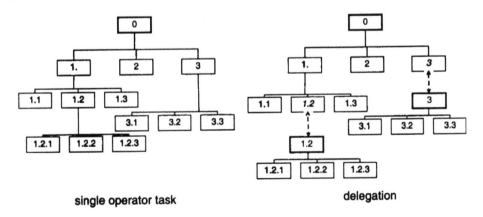

single operator task   delegation

Figure 8.5   *The principle of delegation.*

hierarchy shows the combined task; the right-hand hierarchy illustrates how delegation is enacted. It is important to note that the two hierarchies that emerge always need some modification to ensure that the separate tasks complement one another and the two team members communicate effectively with one another.

The railway example demanded a different solution because of the fluctuating demand created by events that could not be left. Here, a solution was to create team members with general skills and rely on the supervisor to ensure that their combined response was adequate. This sort of solution would entail genuine teamwork and cooperation, with team members being ready to comply with demands and support colleagues in various ways.

The medical example also entailed genuine team skills, where staff need to support one another in ensuring that their own separate efforts created no difficulties for colleagues and that they exchanged relevant information with colleagues about their different tasks. Willing information exchange was critical because different members of the team had access to information that colleagues needed.

All of these examples have training implications where each team member must appreciate the requirements of colleagues.

Assessing workload

Plans in HTA set out how operators must respond to events in order to meet the demands of the task. In discussing plans, Chapters 3 and 4 described how hierarchies of plans which involved contingencies created work demands if several events occurred at once which required the attention of the operator. In the air traffic control task (Figure 6.8), for example, if several events occur at the same time within a sector, the controller must deal with them all. Fortunately, traffic loading can be anticipated and steps are taken to divide sectors between controllers. During busy parts of a day, therefore, two controllers may work one sector, with each attending to different altitudes. As these divisions are anticipated, sufficient radar workstations are provided, together with the technology that allows the radar to be set to display different altitudes. This is a similar arrangement to that discussed in connection with the railway supervision. Through HTA, we can regard workload from two perspectives: contingencies and time-lines.

Contingencies

We need to consider workload in terms of contingencies in circumstances where plans contain conditional statements, such as in the air-traffic control, the railway examples and those examples concerned with system supervision and management. Each of these situations is concerned with the operator monitoring system parameters in order to identify when interventions are necessary to deal with system perturbation. If several things occur together, then several activities will need to be undertaken at the same time. This means that potential workload should be calculated by estimating the likelihood of different, unrelated, events, then calculating their combined probabilities.

When more than one problematic event occurs at the same time, the additional load and attention required of the operator may affect the competence with which each event is carried out. In some cases, these events must be treated together, because the operator must take each problem into account in devising a solution. For example, if the air-traffic controller has to deal with several potential conflicts within an air sector, then the solution must take into account all aircraft within the sector. In other tasks, the solution to problems will be independent. A nurse charged with looking after two babies in a neonatal intensive care ward will deal with their problems separately. If both babies experience a crisis at the same time, then the nurse's problem is one of simple workload, having to deal with two problems at once. In the case of the air-traffic controller, it is inappropriate to separate these interrelated problems, whereas the nurse can be given assistance by someone taking over the care of one of the babies.

Examination of scenarios through time-lines

A common strategy in human factors for examining how operators deal with different scenarios is using a time-line. This is important where there is concern that circumstances may arise with which operators cannot cope. If the consequences of not coping are unacceptable risks, then steps should be taken to share the tasks between two or more people.

A time-line records responses to events as the clock progresses. Figure 8.6 illustrates a time-line. This has been developed on a spreadsheet. Row 1 specifies the time-line against which performance will be assessed. This example deals with how a controller in a railway depot might deal with trains going into and out of service on the main line. The time slot dealt with here is from 6 o'clock to 7 o'clock in the morning. The railway will have been running for a couple of hours already. In the left-hand column in Figure 8.6 in rows 2 to 9 we first see a set of events which may occur and with which the controller is expected to cope. Further down the left-hand column we see the various operations that the controller is required to carry out over the course of his or her duties - on rows 10 to 25. This directly reflects the HTA. It specifies the main goal of bringing trains into and out of service, then how this goal is redescribed. The analyst is at liberty to include any level of detail that he or she considers worth exploring. In this example, it has been decided that two levels of redescription are sufficient. The plans in the HTA are important in identifying the sorts of event that controllers might encounter. Thus, the events in the higher rows all refer to contingencies specified in the plans. The narrow columns to the right deal with successive time-slots. In this example, the time slots are 5-minute intervals. Depending on the nature of the task, the analyst might wish to consider shorter periods - down to a few seconds - or far longer periods.

In situations where operators may be confronted with a variety of contingencies, the analyst can work through different scenarios to evaluate the operator's capability - thus the time-line represents expert judgement. The top 'EVENT' rows show the occurrences that apply to a particular scenario. Row 2 in Figure 8.6 shows the trains that are timetabled to enter service. Train A is scheduled to leave at 0605 hrs, train B at 0620 hrs, train C at 0635 hrs, train D at 0645 hrs and train E at 0645 hrs. Moving down, we encounter some unscheduled events. In row 3, we see that train B fails to move.

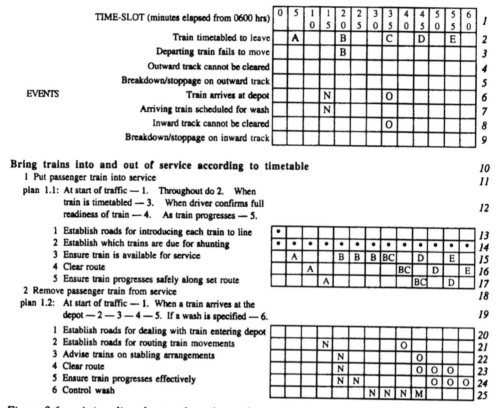

Figure 8.6   A time-line showing how the analyst and task experts judge the coincidence of tasks in a railway control task. The letters refer to different trains in the depot. (The digits in the right-hand column are for reference purposes only to explain the layout of the table.)

Row 6 shows that train N arrives to enter the depot at 0645 hrs and that it is scheduled to be washed (row 7). At 0645 hrs train O arrives at the depot, but the inward track cannot be cleared, so the operator must deal with this problem.

This time-line was used early in the design cycle for the analyst to appraise whether there might be problems with a single person operation. Moving to row 14 we see that, according to the plan, the controller must attend to the timetable throughout in order to establish which trains are due for shunting. This is not an onerous task, but it must be done. The greater risk is that if the operator is distracted he or she may omit doing this. This is a particular problem in hazardous industries where dealing with unscheduled contingencies may cause distraction or work overload and result in the operator failing to monitor a critical parameter.

The analyst may then judge that dealing with the hold-up to train B will occupy attention until it is likely to be cleared, and that B and C can then be dealt with at the same time, as they will occupy different tracks. Then D and E are dealt with according to schedule. Rows 21 to 25 show how the incoming trains N and O are dealt with. If we

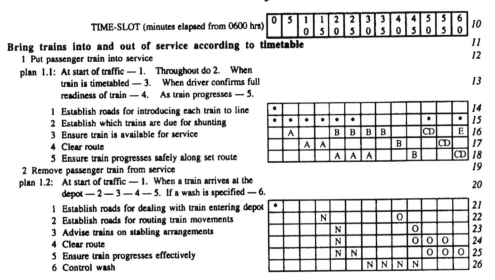

TIME-SLOT (minutes elapsed from 0600 hrs)

Bring trains into and out of service according to timetable

1 Put passenger train into service

plan 1.1:   At start of traffic — 1.   Throughout do 2.   When
train is timetabled — 3.   When driver confirms full
readiness of train — 4.   As train progresses — 5.

1 Establish roads for introducing each train to line
2 Establish which trains are due for shunting
3 Ensure train is available for service
4 Clear route
5 Ensure train progresses safely along set route

2 Remove passenger train from service

plan 1.2:   At start of traffic — 1.   When a train arrives at the
depot — 2 — 3 — 4 — 5.   If a wash is specified — 6.

1 Establish roads for dealing with train entering depot
2 Establish roads for routing train movements
3 Advise trains on stabling arrangements
4 Clear route
5 Ensure train progresses effectively
6 Control wash

Figure 8.7   Timeline recording actual response of controllers.

look up any column we can see the set of operations that are likely to be carried out during any 5-minute period. In the 0635 hrs slot the operator will be dealing with 3 trains plus general monitoring duties; at 0640 hrs, this increases to 4 trains; at 0645, this increases to 5 trains. This should signal a concern to the analyst and prompt discussion with the client concerning the possibility of providing additional staff.

The same basic layout can be use to evaluate actual performance. Thus, it is possible to observe what controllers actually do by recording their actions in a simulator or in the real situation. Figure 8.7 shows a possible controller response to the same events as those indicated in Figure 8.6. Here we see that at 0645 hrs the controller starts to attend to train N and is distracted from attending to train A. So, clearing the route for A is delayed. Train B fails to move (according to the scenario) and attending to train B causes the controller to delay moving A onto the track. All this confusion has caused the controller to forget to monitor which trains are due for shunting, so Train C is overlooked. It is then realised that both C and D need to move, but O must also be attended to, so the controller overlooks E as well - so E leaves the depot late. Even if the analyst failed to convince the client with the evidence from the theoretical examination in Figure 8.6, empirical evidence recorded in Figure 8.7 might be more persuasive.

This example has been used to illustrate the sorts of things that time-lines can be used for. It has not dealt with the fact that some operations are more demanding than others[39]. HTA can make a useful contribution to this sort of investigation because it provides the analyst with raw material from which the time-line can be developed - the *operations* which signal what people are supposed to do at any time and *plans* which specify the contingencies with which they must cope. The principle of a time-line is a simple one and there are many ways in which this basic principle can be used to devise time-lines that may look different and may be used in different ways by the analyst[40] Indeed, a simple time-line was used to represent a time-based plan in Figure 3.8.

The tasks of working within a team

So far, the discussion has focused on the functional aspects of teams and how staff can collaborate by contributing different subgoals and also collaborating and supporting one another, especially by providing information and opportunities for engagement.

Part of the reason why teams operate effectively is that team-members relate to one another in appropriate ways including:

- delegating to colleagues
- handing over at the end of a shift
- accepting problems from colleagues
- controlling the actions of colleagues
- sustaining the effort of colleagues
- motivating and encouraging colleagues
- coaching and instructing.

We can think of these aspects of teamwork as 'team' tasks which can be examined using HTA .

Delegating problems to colleagues

Delegation is a key element in team behaviour. It is particularly important where teams need to be flexible and where a teams promotes personal development of its members. Delegation may arise in teams because of workload or it might feature because one member of the team needs experience in order to develop expertise.

Delegation can only take place sensibly within a framework where the operator delegating first establishes that the colleague is free, able and willing to accept the assignment. This should apply whatever the respective status of the two parties - managers who assign tasks to staff without assessing what they will have to stop doing are not very good at their jobs.

Some work contexts require that the delegator retains responsibility for achieving the goal, while in other work contexts, responsibility can be reassigned. For example, a dentist engaging a locum has not abdicated responsibility for the patient, but has merely chosen an alternate method of delivering the service to the patient. In contrast a small decorating firm unable to fulfil a contract may well arrange for someone else to take over responsibility for that contract. The nature of this sort of delegation should always be made explicit because. Where a person retains responsibility it has implications for their continued monitoring of the work that is being done.

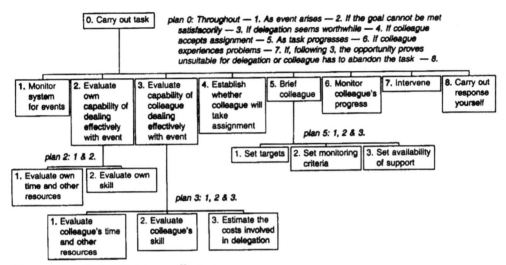

Figure 8.8   *Delegating to a colleague.*

Figure 8.8 shows the HTA of the task of delegating when an operator enlists the support of a colleague to accomplish a goal for which the original operator remains responsible. The analysis is of how the operator carries out the task in broad terms. This entails monitoring events to decide when a response is required, then evaluating whether the operator has the resources to accomplish this successfully without help. If help is required, the operator must check availability, competence, willingness and the extent to which the original operator needs to continue to be engaged, through monitoring

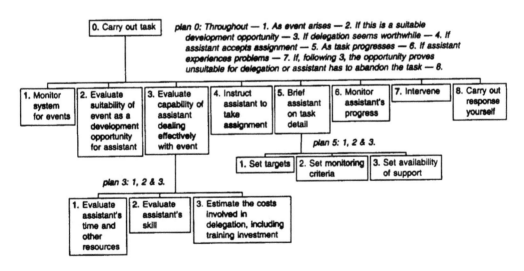

Figure 8.9   *Delegating to a trainee.*

progress, for example. If the operator is unable to secure help from a colleague, then he or she would somehow have to carry out the response anyway (operation 8).

The other main motivation for delegating is where opportunities are sought to enable people undergoing development to gain experience. Again, the original operator is responsible for things getting done. However, when events arise there is an assessment of the suitability of this challenge to the colleague or assistant, then if the challenge is suitable, there are a number of things to do. The assistant's current capability must be assessed. If this is unsatisfactory, then the delegator must assess the effort that will be required to make good this shortfall, for example by arranging some training, and decide whether the effort should be expended on this occasion. If it is decided that the delegation should be made, then the assistant must be briefed, and the assistant's progress must then be monitored to ensure that standards are maintained. Should there be any concern, the delegator may need to intervene. The analysis of this process is shown in Figure 8.9, which is similar in many respects to Figure 8.8.

An example of this process is the training of occupational therapy students undertaking placements with a senior occupational therapist. The senior therapist retains responsibility for the wellbeing of the patient, but must use opportunities presented to extend the experience of the student. A student on a first-year placement will have little experience and expertise upon which to base development. Therefore, the most challenging opportunities will be avoided, except for exceptional students. Occupational therapists have to deal with different people with different problems and personalities at a time of their lives when they are feeling vulnerable. If the senior therapist judges that a particular patient will not appreciate being dealt with by a student, then the opportunity for delegation may be passed over, even though the patient's clinical condition meant that the opportunity would have been suitable to satisfy the student's learning needs. These general considerations apply in all situations of delegation for development. The delegator retains responsibility for the task. If delegation creates too great a risk in a context where the delegator could not intervene and recover the situation, then that opportunity for development may need to be passed over.

Of all 'team' tasks, delegation is, perhaps, the most revealing, because it illustrates a number of implicit skills, decisions and activity that the delegator must be engaged in. By identifying these, the risks entailed in delegation and the skills needed to delegate become explicit.

Shift handover

Shift handover is a common requirement where operating staff must maintain continuity of care or supervision of a system. Examples discussed so far include, supervisory tasks in process control, air traffic control and nursing. In each case, the system under control continues working for a longer period that can be supported by the single shift. If only one operator was involved then, in principle, all the relevant history of the system is known. Where responsibility is shared across shifts, then histories and plans should be communicated to the new shift, ideally in such a way to create seamless control.

In truth, we do not always know what purpose handovers serve. They probably serve both informational and social purposes. The social purposes include helping the worker coming on shift to engage with the task once more and for the worker coming off shift to justify what actions were taken. The informational purposes include providing the new operator with sufficient information and insight about what has been happening so that events can be dealt with more effectively. Thus, a person who has been working for some hours will be able to interpret new events in the light of recent history, while a new worker, unfamiliar with this history, will take some time to establish the same familiarity. Therefore, one role of handover could be to help this process by providing specific information and helping colleagues to orient themselves. To provide useful information at handover entails assessing the needs of the people receiving information and maintaining mental or written records of events judged to be significant during the shift. A handover need not make explicit all relevant information; it may simply provide an outline briefing leaving the newly arrived operator to use records to locate specific information.

Different contexts require different forms of handover. We can contrast handovers in intensive medical care with those in air-traffic control. In intensive medical care, staff have been engaged with one or two patient intensively over their shift, monitoring their vital signs in order to detect any problems. History of a patient over hours is significant. Sometimes recovery can only be gauged over hours or days, and so staff on different shifts need an account of what has been happening. Equally, problems or treatments that the patient has experienced hours previously need to be understood in order to interpret current events. In air-traffic control the events that are significant in interpreting the current situation are comparatively recent. Often, a verbal handover is unnecessary or difficult to accomplish because of attentional demands; the newly arrived operator may need to take time to become acquainted with current events and ongoing plans by reference to current displays and other task information.

Where a verbal handover from one operator to another is warranted, then the general process followed can be represented as in Figure 8.10. This analysis makes it clear that a good shift handover cannot be achieved simply by what is said at the time of the handover. In systems where people must share responsibility across shifts, operators must be conscientious in collecting information of relevance to their colleagues as the shift progresses. This information is recorded in a shift log or committed to memory.

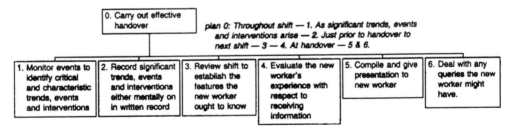

Figure 8.10   Elements of shift handover.

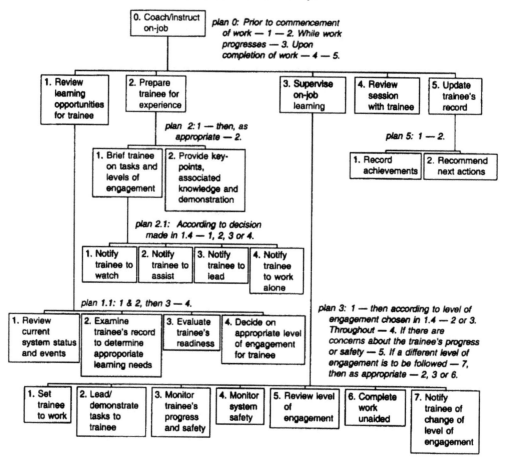

Figure 8.11   Coaching and instructing.

Coaching and instructing

A common team skill is coaching and instructing. This is where an experienced team member assists a new team member to learn how to carry out operations. In some contexts this is called *staff development* and in some contexts it is called *on-job instruction*. The common thread is that a novice is required to learn operational skills whilst the normal activities of the task are taking place. Sometimes this is done because it is cheaper than alternative methods of training and sometimes it is done because it is the only way in which the new starter can experience the necessary detail and rhythm of real work events. In any case, in all training there has to be an occasion when a new starter experiences the operational situation for the first time and it is advisable that they do this under the guidance of a more experienced colleague.

The intensity of coaching activities may vary. *Staff development*, for example generally takes place over time with the trainee being given time to accomplish a set of activities to be reviewed later on, whereas *on-job instruction* is usually seen as more immediate, with an instructor attending to the trainee as events take place. Despite this, the general activities of coaching and instruction can be modelled using HTA, as in Figure 8.11, to reveal various critical steps that need to be accomplished effectively.

Controlling the actions of colleagues

There are many situations where operators working in different areas are obliged to cooperate with one another because each has a different view on proceedings. A common situation is where maintenance or process staff working outside in an industrial system are reliant upon the collaboration of colleagues in a control room. The outside tasks might be reliant upon information that is only available through instruments in the control room. The control room operator might be able to provide the colleague working outside with information about current parameters of the system under investigation, or control parts of the system to enable safe access or make tests, or to coordinate activity with that of other people working within the system. Where these arrangements occur, this can be dealt with as a kind of *delegation*, in the same way that the packaging line operator worked with the maintenance technician.

Exchanging task information with colleagues

Another team task is concerned with each operator working in a way to cause least disruption to colleagues. If doctors administer treatments in a way that causes the nurse extra work, then, rightly, this is inappropriate if there is a less disruptive option. Where staff leave photocopiers jammed for colleagues, then the colleague will be disrupted. Team members may identify information of value to colleagues. If they are able to judge the criticality of the information with regard to what the colleague is currently doing, they may be able to wait for a less disruptive time to convey this information.

This sort of team interaction characterises an effective team member. By identifying this team requirement it is clear that team members need to understand what their colleagues are trying to do rather than merely training people in their individual skills.

Concluding remarks

Understanding how people work in teams and the jobs that team members must individually fulfil to meet the team's goals is crucial to the success of many organisations. Once teamwork is considered, attention is often directed towards social skills. However, there are also important functional aspects and this is where HTA can contribute.

This chapter has demonstrated three main uses to which HTA can be put in helping to understand team behaviour. First, the hierarchical organisation of the HTA lends itself to a helpful understanding of how people must interact in order to accomplish the team's functions. Different contexts warrant different approaches. In some situations, tasks are partitioned between several team members on the basis of their respective workloading or their respective skills. This can result in different people doing different things within the team. In this case it is necessary to understand how the tasks of each team member need to be adapted to ensure proper support - usually through the medium of providing information that colleagues can use to accomplish their parts of the joint activity. In other situations, team members may be doing the same things as each other, in which case, appropriate supervision is necessary to ensure that all events are suitably dealt with.

The second way in which HTA can contribute is in providing a basis for the analyst to assess workload. In many cases, workload arises because operators are required to deal with several events simultaneously. They may become overloaded and make mistakes, or they may simply become distracted and fail to attend to certain duties. By considering the plans within the HTA, the analyst is able to identify those events that might coincide to contribute to workload. It then depends on estimates or evidence that such events will coincide to determine whether the operator may be excessively loaded. Another method for assessing workload is using time-line analysis. Here, HTA is useful in providing the operations that people must carry out over time. The plans in HTA set out the range of events with which operators may need to cope. In this way, different scenarios can be evaluated.

The third role of HTA within team activities is in helping analyse the ways in which people must interact when operating within teams. Team members are engaged in delegating, supporting, instruction and providing each other with information, especially as responsibility is passed between shifts. These methods of interaction can themselves be analysed using HTA to reveal the sorts of skill that good team members must possess.

Chapter 9

Information and skill

Information is fundamental to skilled performance and it is important that designers represent information and provide control in a form which the human operator is able to use.

This chapter discusses the importance of information and control and shows how these ideas are central to the HTA. The consequences of inadequate information are discussed.

The chapter then focuses on the issue of information requirements specification to illustrate how HTA can be used to identify information through the examination of both operations and plans.

Finally, benefits that HTA can offer the representation of information are illustrated.

Introduction

This chapter will consider an important aspect of redesigning the task, namely how the presentation of information and controls affects performance. HTA provides a framework in which each aspect of the task can be considered in turn, enabling the analyst to consider whether information and controls are sufficient to support effective performance.

Information is that which an operator uses to distinguish between different states of the world. This may be achieved using an artifact that has been deliberately designed to provide information, such as an indicator or a trend recorder, or it may entail something that varies in the environment in a significant way that the operator learns to understand. Thus, a speedometer provides the driver with an accurate measure of speed and this is used by the driver to regulate how fast the car is going. The view of the road shows the driver the direction of the car and the state of the traffic provides cues to indicate how traffic can be negotiated. The view of the road and the state of traffic are both information sources that the driver uses to regulate aspects of driving performance, but neither are artifacts that have been explicitly designed for this purpose. Experience teaches the driver how to recognise different situations requiring different responses.

Controls enable the operator to make changes to the system being controlled. Activating controls entails muscular action and physiological processes. Here, too, information is crucial. Information helps the operator determine the location of the appropriate control. It can also indicate how the control should be activated, thereby reducing the amount the operator needs to remember. For example, a switch which is clearly labelled to inform the operator that 'down' means 'on' and 'up' means 'off' is easier for someone to use with confidence, than one which is not labelled. Where there is no such label, the operator has to remember these positions or operate through trial and error - and this may be inappropriate if a wrong action cannot be recovered.

In the design of controls there are many aspects concerned with physical action and effort. For example, the size, location and effort that needs to be exerted to activate a lever must be within the physical capability of the operator. The size of knobs and latches must be such that the operator who is to use them is capable of their manipulation. Environmental conditions will also influence the operator's fatigue and recovery as a continual effort is required. These non-informational issues are crucial and are the subject of disciplines such as anthropometry and environmental ergonomics[41]. HTA is relevant to such issues to the extent that the need to utilise a control is flagged through the task analysis. If the need to use a control requiring a certain degree of exertion is identified, then this can justify further investigation of the ergonomics involved. However, the main benefit of HTA with regard to helping the analyst better understand the place in which people are working, is in identifying the operator's information needs.

The importance of information in understanding human performance and the controls through which the operator is able to carry out action was described in Figure 1.4. HTA provides a routine way of examining information and action. As the analysis progresses and the analyst judges which parts of the task warrant attention, so consideration can be given to whether the facilities providing information and control are best suited to supporting human performance. For example, an operation may show that the operator must read or listen for crucial information. The analyst must be satisfied that suitable signals can be seen or heard by the operator in the operating context. To this end, the analyst must be alert to any feature that masks or prevents this information being detected or discriminated. If an operation depends upon actions where energy is expended, then knowledge of exercise and fatigue will apply, as will knowledge of the thermal environment and how it influences exhaustion, discomfort and recovery. If action depends upon movement, especially within a constrained space either because the space limits movement or because there are limitations on how controls are arranged, then knowledge of anthropometry will be useful.

When occupational tasks were mainly carried out manually, they entailed perceptual motor skills where the operator or artisan directly fashioned raw materials. To work leather or wood entailed working directly with leather or wood, sensing it through sight, touch, smell and feel, and manipulating it directly albeit, with tools designed to make operations easier. This meant that most reliance was placed on the artisan acquiring skill in order to work effectively, although other ergonomics issues, such as a suitable environment in which to work, was also vital.

Increasingly, people are employed to interact with the systems they operate through the medium of a man-made interface. Interfaces were often a set of instruments, knobs and levers connected to the system by mechanical linkages or pneumatic devices. The design of such interfaces was constrained by the physical characteristics of the display and control devices and the constraints on the mechanical or pneumatic links to the processes being controlled. A great deal of ergonomics has been directed towards making these interfaces as effective as possible. Designed well, such interfaces can make the task easier; designed badly, interfaces can make some systems inoperable, or at least limit the effectiveness of performance. With the widespread introduction of computers into the workplace, there is far greater freedom for the designer to represent the system in different ways with less constraint than hitherto. Presented with this greater freedom of design, there is an opportunity to get it right and also the opportunity to get it wrong. To be effective, the designer needs strategies to ensure that appropriate design decisions are taken.

The reliance on information in tasks - some examples

Every example of HTA in this book emphasises the importance of information and control. Figure 3.1 illustrated the task of saving text in a wordprocessor. This was a straightforward fixed sequence plan. None the less, information is crucial in providing the operator with information to control cursor movement, across the screen and to confirm when a required menu choice had been made. This information is provided by screen cues that the interface designer has included. Without this information, the task could not be done.

Boiling an egg, shown in Figure 3.2, entails a contingent sequence plan which depends upon recognising that water is boiling and then that sufficient time has elapsed.

If the operator cannot judge when boiling commences or is unable to measure the passage of time, then the egg is unlikely to be satisfactory. Distraction elsewhere will affect judging when boiling starts and absence of a reliable timer will affect judgement of elapsed time.

In the batch processing task, shown in Figures 6.3 to 6.5, success depends upon knowing when things have happened, when states have been achieved, and how to set values. Thus, if sight-glasses through which the product is observed are dirty, or there is insufficient illumination to make a reliable judgement, then errors can easily be made. This task was also discussed in Chapter 7 and the plant layout shown in Figure 7.3. A key aspect here is that the demands of management caused the operators to be away from the location where they could, routinely, monitor the distillation rate. As a consequence, operators fell back on a conservative strategy where they limited the distillation rate to one they could maintain without risking the batch. This accounted for excessively long batch times. The problems emerged, in part, through failing to provide adequate information in the correct place.

In stabilising the distillation train in Figure 3.8, the operator was required to stabilise one distillation column, then maintain this at a steady state while the next column was added. Information is crucial to each stage of stabilisation and the operator must know the various temperatures, pressures, levels and flows in the vessels in order to make adjustments to steam and flow rates. If straightforward information is unavailable, then the operator must infer values from the information that can be observed, which increases the cognitive loading on the task. Moreover, the earlier columns must be continually monitored as the later columns are stabilised. With no key parameters to show the operator that all is well with the earlier columns, the operator must either engage in more complex reasoning about the status of units upstream, which adds to workload, or make a risky judgement from observing fewer parameters which do not provide exhaustive evidence about the earlier column.

The chlorine balancing task developed in Figures 3.14 to 3.19 is wholly dependent upon information about events that have occurred and the status of other parts of the system. If information is unavailable, the task simply cannot be done. If information is only available at a cost, then this can have other adverse effects on the system. For example if the current status of the cell units can only be obtained by visiting the cell-rooms or telephoning the cell-room operator, then decision times will be delayed while the enquiry is made. Moreover, if the attention of cell-room operators is occupied with the enquiry from the chlorine controller, this reduces their response to events on their own plant. Often, in such tasks, risks are taken if information is not readily available. In the chlorine task, the controller who finds information difficult to obtain may assume that a resource previously available is still available, because he or she can recall no evidence that its status has changed. If parameters are easy to confirm, this risk is less likely to be taken. Taking such a risk could result in inefficient decisions or accidents being caused.

Further analysis of the task of maintaining best service in the underground railway supervision task shown in Figure 4.6 identifies the need for the supervisor to make decisions concerning gaps in the service. This may be done by delaying trains, speeding them up, or introducing more trains into service. Making such decisions involves a number of considerations. One requirement is to avoid congestion at stations. Deciding whether a new train is required to be brought into service entails judging the extent of the congestion on the station and the current loading of the trains in service. If the supervisor is provided with video information and has knowledge about the kind of events that could cause overcrowding on certain stations at certain times, then he or she can make judgements of congestion without resorting to contacting local station staff who may be otherwise engaged. Moreover, if current train loadings can be conveyed directly to the supervisor, the capacity of existing services can be judged more readily. Hence, there are trends towards the use of video on stations and weight-sensing devices on the bearings of trains to provide this information directly. Generally, by understanding what information the operator needs in order to make a decision, the designer knows which steps should be taken to ensure that information is satisfactory.

The ultrasonic inspection task, described in Figures 4.4 and 4.5 is also awash with information issues. Operators need to know which probes are required and how to distinguish between them. They require information to help them control the direction and pressure of probes. They also require system feedback to show the reflection of ultrasonic signals in order to judge whether there is a flaw in the object being inspected. If any of this information is missing, then performance will suffer.

Characteristic problems with information and control

Three common sorts of information problem are *degraded feedback, breaking the information and control loop* and *information fragmentation*

Degraded feedback

If task feedback is unsatisfactory in providing reliable information about the state of a system, then control actions are unlikely to be effective. A problem for bus travellers or theatre-goers arises if distributed booking agencies cannot view an up-to-date central record of what seats have already been booked. If the agency sells seats in good faith, and the communication to the central database is not immediate, then double-booking may occur. Such problems require a technical and a procedural solution.

There are other situations where the problem is concerned with the honesty and objectivity of colleagues. Increasingly, public service agencies are being judged according to performance indicators. This risks the agencies distorting their reports concerning how their systems are performing. Social-work managers and head-teachers, anxious to report favourably to their masters, may paint a rosy picture of their organisation's achievements, even where they are underachieving. If these positive reports are then used to divert resources to more 'deserving' cases then the underachieving organisations are likely to fare even worse. The same problems occur in organisations with regard to safety reporting systems. If staff are reluctant to report 'near-misses', then management may adopt a rosier view of conditions than is justified. Other systems suffer where one group of people are responsible for collecting information that another group must use. Often, management services departments require information for planning purposes that other staff must collect. If other staff are reluctant or unable to collect this information then subsequent planning will suffer.

In industrial systems, engineers may fail to provide straightforward and reliable means for operators to judge the outcome of actions. If there is no clear method for displaying a parameter, the operator will need to draw inferences from those parameters that are measured. Degraded feedback problems arise where the feedback upon which the operator is relying does not match the real state of the system.

Breaking the information and control loop

'Breaking the loop' problems arise where it is not possible for operators to view the necessary control information in conjunction with the controls they are supposed to operate. This can also arise when task conditions affect the operator's willingness or ability to collect this information. The operator may make assumptions about whether the system's current status is satisfactory. This is risky behaviour, which could have a serious outcome if there was anything wrong with the system.

Numerous examples of this sort of problem can be seen. The batch plant layout shown in Figure 7.3 created problems because task conditions discouraged operators from frequent visits to where information was displayed, with the result that they operated an over-conservative strategy. Similar problems can be seen in control rooms where instruments and associated controls are placed distant from each other on different walls. The consequence is that controls cannot be adjusted continuously according to the target parameter being adjusted.

Such problems are not simply the province of old-fashioned working environments. Instances are also frequently observed in systems where staff are employed to supervise complex systems through the medium of computer interfaces. One consequence of information technology is that information can be transmitted and displayed more cheaply that hitherto. In earlier display technologies, sensing devices were complex and costly and the pneumatic or electromagnetic systems used to convey information required a separate channel for each item of information being transmitted. With computers, sensing devices can collect more information locally, then send this to remote control-room computers via a central data bus which is cheaper to install and maintain. While process information used to be at a premium, with new technology it far less expensive to convey. Also, while earlier display technologies relied on dedicated instruments set out on control panels for each item of information conveyed, modern displays rely on visual display terminals (VDTs) which can represent lots of information in different ways. Despite their obvious advantages, computer screens are small compared to wall mounted control panels. Thus, more information than before has to be displayed on a smaller area. The solution is to enable the user to access numerous screens each displaying different subsets of information. The designer's problem is to decide which information to display on which screen and how to help the operator to navigate between screens.

Faced with the task of organising such systems, designers look for strategies to guide them. In many cases, the designers instinct is to use a *geographical* basis for organising information. The screens selected can represent different geographical or

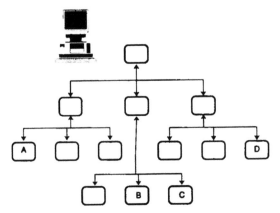

Figure 9.1   A hierarchical menu structure.

spatial areas of the system. Thus, in Figure 9.1, the top screen would represent the whole system, while the next level down represents areas of the system and the next level down represents areas of these areas. Geography is not the only way of organising information. In some nuclear plants, information is grouped according to different sorts of functional measurement - one set of screens represents *flows* through pipes, another *levels* in vessels, another *temperatures*, another *pressures* and yet another *radioactivity*. Thus, the operator needs to travel to different screens to obtain information about different sorts of parameter.

These are convenient solutions for a designer because every item of information has its place and both designers and operators can share the basis on which the information is organised. Thus, operators need to be told the principles of organisation so that they will know, in theory, where each item of information is located. This arrangement is fine to enable people to retrieve specific items of information but this sort of solution does not reflect how the information will be used to carry out many tasks. In control tasks, these arrangements can result in breaking the loop. For example, it might be necessary to view screen A (in Figure 9.1) to judge the state of a target process parameter in order to control it from a controller in screen B. Having to work through different screens becomes cumbersome, especially where a decision must be taken quickly. The operator might choose to rely on memory of the state of one or more of these parameters and assume that circumstances are unlikely to have changed since they were last viewed. Alternatively, because the operator cannot be in two places at once, he or she may choose to work in tandem with a colleague looking at different parts of the same control system through different visual display terminals. This is very inefficient from the perspective of maintaining control and very costly in terms of manpower. It is far better to develop menu systems in accordance with the *information requirements* of operators and to use these insights as a basis for generating *operational* menu systems. In this way, information and control devices that must be used together can also be accessed at the same time. Many computer systems are designed to provide such flexibility, but many are not.

Breaking the loop problems are observed in systems that employ the latest display technology as well as in more old-fashioned systems. They represent a failure by engineers and designers to anticipate the information requirements of staff carrying out their tasks. We shall shortly discuss how HTA can be used to clarify these problems. *Information fragmentation* problems, which will be discussed next, can also occur with these information arrangements.

Information fragmentation

In carrying out planning and decision-making tasks, operators must often examine several items of information in order to draw appropriate inferences to enable them to select appropriate control actions. This is required in all problem solving activities, such as diagnosis, recovering target operating conditions, compensating for disturbances, or changing operating characteristics. In such tasks, choice of action is rarely straightforward. Effective performance depends upon operator skills, but these skills depend upon being provided with appropriate information. If this information is not available, the resultant decisions will be suboptimal or wrong. The same applies if the information is not easily accessed. One consequence is that operators under stress may make assumptions, rather than carry out protracted procedures to access all the necessary information from the computer. In the chlorine balancing task, described in Chapter 3, operators needed current state information to make decisions on rebalancing. However, there was the risk that operators would rely on their memory of system states when making decisions and fail to establish the current system status.

Another illustration of this problem emerged in a plant which was composed of three main interacting areas each under the supervision of a different process operator. A computer-based control system was being developed to replace an existing conventional control panel. Figure 9.2(a) illustrates the complexity of this plant. While the main feed was to Area A and the main product was taken from area C, there were many complex interactions between the three plant areas. These were in terms of heat exchange, automatic control and various by-products which were fed to other parts of the plant to enable processing. This meant that if a problem arose in any of these areas of plant, its effects would be referred to each of the other areas of plant. This would

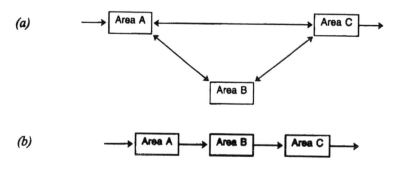

Figure 9.2   Flows of information, energy and product through plant: (a) complex arrangement if interactions; (b) simple arrangement of flows.

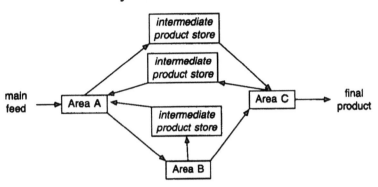

Figure 9.3   The arrangement of intermediate product stores in the system.

affect fault diagnosis strategy, which can be seen by comparing the arrangement in Figure 9.2 (a) with that in (b). In Figure 9.2 (b) the flow is simple - product, energy and information all flow in one direction. A problem in A would refer to B and C and then to the final product. A problem in B would be referred to C and the final product, but it would not affect A. A problem in C would affect the final product but not A or B. In systems such a (b), therefore, it is relatively easy to isolate the part of the plant in which a problem resides. In (a) however, this is far from straightforward.

One consequence of this complexity was that operators were not expected to solve problems immediately, but first had to try to work out how the plant conditions could be readjusted to optimise production while fault-finding took place. Typically, this required operators to decide whether it was possible to continue production by using intermediate stores of by-products to keep various parts of the process functioning. Figure 9.3 represents Figure 9.2 (a) in greater detail. The main feed was first processed through Area A. From Area A, the main product stream fed to Area B, but a by-product was also produced which was stored to contribute to the processing in Area C. Equally, Areas B and C created by-products which were used by area A. Thus, if there was a problem with, say, Area B, this might affect the processing of Area A, which would then affect the capability of Area A to produce the intermediate product to contribute to Area C. So, if the operator was able keep area A working, then Area C could also continue to function and complete the manufacture of product that was in the system. This was possible if the operator was able to maintain stocks of the various intermediate by-products. Indeed, the operator's main job was to make these judgements in order to minimise the effects of the problems that might arise. To do this, the operator had to examine information from various parts of the plant. Thus, if Area A showed a problem, then intermediate stocks serving A, B and C had to be inspected to see whether compensation was possible.

Figure 9.4 represents how the information was originally represented to operators on conventional control panels. All of the information from Area A was represented on control panel A, Area B was shown on control panel B, and Area C on panel C. To make compensation judgements, the operator had to move around the control room to view the information represented on all panels.

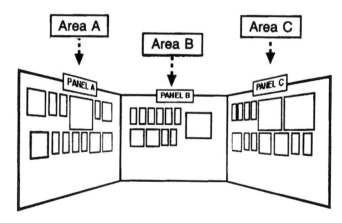

Figure 9.4   Layout of control panels

The new computer-based display system was laid out as shown in Figure 9.5. Again, information from each of the three plant areas was represented via a different console. The client assumed that since all that was changing was the representation of information, existing staff could adapt easily to the new arrangement. All that was needed, it was assumed, were some training exercises whereby the operators would be presented with problem scenarios through which they could become used to the new form of information presentation. To design these exercises, the analyst examined some representative problems to establish scenarios with which the operators could practice adapting their existing high level of skill to the new arrangement. However, through applying HTA it soon became clear that existing skills would not transfer so easily and that the new arrangement contained some major problems.

With the existing control panels, operators were able to scan information that was presented simultaneously and even look for patterns on the displays. Moreover, the operators responsible for looking after the three different areas could work in harmony with each other, because they each had access to all of the information presented on the

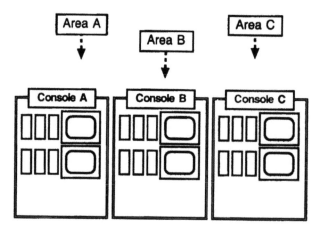

Figure 9.5   Layout of computer control consoles.

control panels. With the new arrangement, operators had to locate each item of information by working through a hierarchy of screens. Thus, the Area A operator would examine information about Area A by working through a hierarchy of screens on console A. But if the problem also required information from console B, then operator A, would have to go to B, then interrogate the menu hierarchy and then to C, and possibly back to B, then back to A. Not only was every decision protracted in this way, there were frequent clashes between the three operators who also needed to examine their respective screens. Analysis of the tasks using this arrangement predicted complete chaos.

Once this problem was recognised it was, fortunately, relatively easy to deal with. By using the HTA to examine the procedures followed to make typical decisions it was possible to identify which information would need to be considered by the operator under different circumstances irrespective of its location within the overall plant. The problem could be dealt with efficiently by reconfiguring the displays to group information *operationally* and not simply *geographically*. In this way, operators could go to any one of the three consoles and call up 'operational' screens which contained · all of the information necessary for each respective operating decision. By following this design principle, the need to move between screens to make decisions was minimised. The geographical arrangement was retained to enable retrieval of isolated items of information. This meant that scenarios that had not been identified could still be dealt with albeit inefficiently. These could then be added to the operational set if it was felt to be worthwhile.

This general solution meant that operator training could now proceed through providing practice with scenarios, because the representation of information was no longer fragmented and there were no longer cumbersome procedures necessary in order to obtain each item of information.

Identifying information requirements in operations

Information requirements specification is the systematic process of examining a task in order to determine the information and other facilities that a designer should provide at an interface to support operator performance[42]. This process can be illustrated by reference to a simple operation - switching on a light. This requires a device (a switch) to enable the action. Then the operator needs the following information:

- the identity of the light switch and where it is located - so that it can be found
- how the device is used - so it can be switched 'on' or 'off'
- action feedback - to indicate to the operator that the control is being activated in the manner required
- system feedback - to show the consequences of the action on the system under control.

A designer who ensures that these sorts of features are apparent in the design is providing help for the operator carrying out the task. The designer may make *identity and location* clear for the operator by effective labelling and positioning of controls. If the designer fails to make the identity and location of the light switch clear, then the operator, confronted with several switches, will need to learn which one does which job, or rely on trial and error. In the design of many domestic devices or items of electrical office equipment, aesthetics are preferred to clear functional design, with on - off switches cunningly disguised to improve the smoothness of line of the object. In these cases, the user has to get out the user-manual to locate these switches.

Good design can also show how a device is *used*. One aspect of this has been called 'affordance'. This is where the representation of the device helps to suggest to the operator what needs to be done to achieve a desired effect. Instances of good affordance are seen on interior swing doors, where door-plates tell a person that a door should be pushed, while a handle suggests pulling or turning. Some furniture design deliberately disguises the purpose of the object. For example, many hotels seem to think that their customers are impressed by disguising the refrigerator as a chest of drawers - it is only when one is looking for the Gideon Bible that the cache of whisky and gin miniatures are finally discovered. Another aspect of good design to help guide a person in using a device is the use of conventions or 'population stereotypes'. If the designer appreciates that operators will always assume that one sort of control action will yield a desired outcome, then including this in design is most sensible. Thus, people may assume that turning a knob clockwise will increase the value of a parameter being controlled; putting a switch to the 'down' position will suggest to people in some parts of the world that a device will be switched on, whereas 'up' switches it off. It is important that a designer appreciates the assumptions a user or operator population will make about the relationship between controls and displays and which population stereotypes apply.

If there is unsatisfactory *action feedback* the operator cannot be certain that the device is working. A keyboard operator who does not receive routine confirmation from the pressure exerted that keys are being depressed, must keep looking at the screen to confirm that typing is being effective. The same material may be processed successfully, but at a much slower rate and with more errors requiring correction. Finally, *system feedback* confirms that the right things are being done. If there is prompt and appropriate system feedback the operator will more quickly determine whether the action selected was appropriate in dealing with a current situation.

All of this means that if such an operation is identified in an HTA - and there are many such operations in tasks - then the analyst can list the sorts of information that needs to be present for the task to be carried out effectively. This information can be used by an interface designer and a training designer to ensure that a suitable blend of design is adopted.

Types of operation

The example just discussed dealt with supporting the operator carrying out a simple action. As HTA is developed, it is often possible to understand the task in terms of similar types of operation. Most operations encountered in this way can be categorised as one of the following four main types:

- actions
- communications
- monitoring
- decision-making.

Actions

In carrying out an action, the operator affects change to a system, such as switching on a light, starting a car and moving a computer mouse. We can specify three different types of objective with regard to identifying information requirements:

- activating or deactivating equipment
- preparing or setting equipment
- adjusting equipment.

Activating or deactivation equipment is when the operator takes an action to start or stop something. This is usually done through a knob, a key, a switch or a lever. Starting a car or booting up a personal computer are obvious examples.

Preparing or setting equipment is undertaken at the start of longer processes when it is necessary to ensure that systems are in a proper state of readiness when they are called on to be activated. Cooking, surgical operations and maintenance each require such preparation. If these things are not done then the main operation will fail or be delayed, and this can be critical. For example, if an item of emergency equipment is not available when required and in a state to be operated then, in any subsequent emergency, the system is likely to deteriorate further.

Adjusting equipment is done when a parameter needs to be regulated. For example, changing the volume on a television set or adjusting the speed or direction of a motor car. Such adjustments are commonplace in using many devices.

Communications

Colleagues working together within a team must often share information because different team members have access to information that may not be directly available to colleagues. So this information must be *communicated*. Communication is a form of action because it requires the operator to expend energy in speaking, writing, reading and listening. It influences aspects of the system under control by redistributing system knowledge so that colleagues are able to regulate their actions with information which they did not obtain directly. Four such operations are:

- record information
- give instruction
- receive instruction
- obtain information.

Recording information is important in situations where the operator must record an account of activity for auditing purposes, or where crucial information must be identified and recorded for colleagues to use. The medium through which this is done is either a computer system, a written log or through memory. To record information in each of these it is necessary to identify the information to be recorded and to locate where the record should be made. Processes of typing, writing or speaking into a recording device all entail actions that need appropriate forms of feedback to ensure they are carried out properly.

Giving instruction is where one person communicates information to another. In such communication, it is important that the sender makes clear the topic of the message, who is sending it and the content of the message. It is also vital that the sender is satisfied that the message is transmitted as required and that it is received successfully. In many safety-critical situations, including air-traffic control, medical and military contexts, communications are formalised. Standard codes are used to identify sender and receiver, terminology is standardised and restricted to avoid confusion and formal confirmation of messages received is given by the receiver.

Receiving instruction is necessary in situations where the receiving operator's plans depend upon someone else specifying which operations the receiver must carry out or the task criteria that must be observed. Typically, the receiving operator's task entails *plans* which state that operations must be selected according to instructions given by another named person. The person issuing the instruction may be in authority, such as a manager or supervisor. Alternatively, the person giving the instruction may be a colleague who is in a better position than the receiver to know when or how something should be done. Thus, a person helping a colleague to park in a tight space is in a better position than the driver to know when to stop. When two people are lifting a heavy object, it makes most sense for one of them to tell the other when to lift and when to lower, so that accidents can be avoided. A control room operator will collaborate with a maintenance engineer working on site to tell the engineer when some things must be done, while the engineer, in turn, tells the operator when some other things must be done. None of these examples necessarily implies a fixed authority relationship. They merely show that colleagues must often furnish one another with information to enable each to carry out their tasks. Hence *receiving instruction* complements *giving instruction*.

Obtaining information whether through verbal communication or from written records is the complement of *recording information*. Whereas *receiving instruction* is concerned with operators changing their goals or criteria for performance according to a message passed from someone authorised, *obtaining information* is simply concerned with establishing the values of given parameters to enable subsequent decisions to be made. For example, the nurse making a judgement about how best to feed a baby will take account of the baby's previous history from records which indicate the baby's current health, previous rate of growth and information about the baby's previous feeding behaviour. The nurse still makes the current decision about feeding, but this decision will be better informed by the information provided through records and from colleagues.

Monitoring

Monitoring behaviour has been referred to repeatedly throughout this book. It is an important activity in maintaining system states and carrying out control actions. Three types of monitoring are distinguished:

- monitoring steady state or rate
- monitoring to anticipate a target
- inspecting equipment.

Monitoring steady state or rate is where the operator has a target with which a process parameter must comply. Maintaining a car speed at 45 kilometres per hour entails recalling this target value then continually comparing the car's actual speed with this target. Monitoring behaviour is only ever important if deviation from target must be dealt with. If the car's speed exceeds 45 k.p.h., then the driver must take an action to reduce the speed; if the speed drops, then the driver must speed the car up. The actions of slowing or speeding the car up are dealt with as action elements. So, in the context of an HTA the analyst will analyse 'maintain car speed' and then identify two different operations where the operator uses the interface for different, but related, purposes.

Monitoring behaviour might be concerned with comparing an aspect of the *current state* of a system with a target, or it might be concerned with comparing the *rate of change* of a parameter with a target rate of change as in the batch plant described in Chapter 7. That is, the operator monitors the rate at which a parameter is increasing and acts to increase or decrease the rate if it falls outside of a target. To support monitoring steady state or rate, the operator must be told the parameter to monitor, must be told the targets against which to monitor and must be shown the current value of the parameter being monitored.

Monitoring to anticipate a target is where change is required when a certain value is obtained. If a vessel is heating up, it might be necessary for the operator to identify when a particular value is reached as a cue for the next action. This is a form of monitoring in that a parameter must be observed and compared with a target value. When the observed parameter matches this target value, the next action is cued. Thus, in *monitoring to anticipate a target* it is expected that the target value will be reached, whereas in *monitoring steady state or rate* the operator is not necessarily expecting the parameter to go outside of tolerance. *Monitoring to anticipate a target* only needs close attention as the parameter approaches the target value. Making toffee is a good domestic example of this. Sugar must be heated up until a temperature is reached where the boiled sugar turns to toffee of the required consistency. This can take a long time and there is no need to monitor the temperature or the state of the boiled sugar continuously. However, when the target temperature is reached it must be detected swiftly and the mixture removed from the hotplate. If this is delayed, then the mix will overheat and the toffee will change its characteristics, become more brittle and possibly burn. This means that the skilled operator will know the targets to be monitored for a successful outcome, but will also know the delays that can be assumed before intensive monitoring is required.

Inspecting equipment entails the operator judging whether an item of equipment is currently fit for its purpose. An inspection task involves a number of procedures in which parts of the device being inspected are revealed for inspection to take place. This is followed by a number of acts of inspection in which parts are judged to be satisfactory or deficient. If they are deficient, then remedial action takes place. The inspection phase entails the operator judging the state of some aspect of the device against a criterion or standard. The operator must know what feature is being inspected, the current state of this feature and against what standard it should be judged.

Common information requirements

If we put these different operations together, we can summarise the sorts of information required by the different types. This is shown in Table 9.1. Table 9.1 is complemented by Table 9.2 which sets out descriptions of the different information types.

Decision-making

The final operational type, listed earlier, was *decision-making*. This includes diagnosis and planning. These operations have been encountered extensively throughout the HTAs presented in this book. Establishing information requirements for decision-making operations is less straightforward that dealing with the other sorts of operation discussed.

Diagnosis is concerned with establishing how to return an unacceptable operating situation to an acceptable one. It might involve determining which of several potential causes of a problem has actually arisen, such that the problem can be resolved and the system returned to a healthy state. It might simply involve deciding on a course of remedial action to return the system to a healthy state without worrying, unduly, about the cause of the problem. In many mechanical or electronic systems, diagnosis entails deciding which item of equipment has failed so that it can be replaced or repaired. Thus, identifying the faulty item specifies where the repair must be carried out. In many systems, however, formal identification of the problem is not necessary or even possible. Thus, medical staff are concerned to identify a regime through which the patient will recover and this does not always mean that the problem must first be identified - doctors and nurses will decide what treatment seems to fit and will then monitor the patient's response to determine whether the treatment should continue, whether it should be modified and when the treatment can be withdrawn. Sometimes information is used in diagnosis to reason about alternatives, for example, to enable diagnostic rules to be used to make inferences about what has gone wrong or to eliminate possibilities which are inconsistent with information that is known about the task.

Table 9.1   *Typical information requirements for different operational types.*

Information requirements

	Identity & location of control	Method of control	Identity/location of target param.	Target value	Band-width	Current status of parameter	Current status of control	Wait period	Action feedback	Control feedback	System feedback
Actions											
a1 activating/ deactivating	✔	✔	✔	✔		✔	✔		✔	✔	✔
a2 preparing or setting equipment	✔	✔	✔	✔		✔	✔		✔	✔	✔
a3 adjusting equipment	✔	✔	✔	✔	✔	✔	✔		✔	✔	✔
Communications											
c1 record information	✔	✔							✔	✔	✔
c2 give instruction	✔	✔							✔	✔	✔
c3 receive instruction	✔										
c4 read information	✔										
Monitoring											
m1 monitoring steady state or rate	✔		✔	✔	✔	✔					
m2 monitoring to anticipate a target	✔		✔	✔	✔	✔		✔			
m3 inspecting equipment	✔		✔	✔	✔	✔					

Table 9.2   *Description of information types.*

Information type	The function of Information
• Identity & location of control	Where the item to be observed can be found.
• Method of control	How the control device should be manipulated to achieve different control outcomes.
• Identity and location of target parameter	Where the target parameter is located.
• Target value	The value the target parameter should assume.
• Band-width	Upper and lower limits for the target parameter (where appropriate).
• Current status of parameter	The current value of the target parameter.
• Current status of control	The position at which the control is currently set.
• Wait period	How long before attending to the target parameter.
• Action feedback	How the control is being manipulated.
• Control feedback	Whether the control is activated.
• System feedback	What effect the control has on the system.

Sometimes information is used to test hypotheses - once a hypothesis has been developed, the diagnostician can compare what should be observed on the assumption that the hypothesis is true with what is actually happening.

Planning is concerned with deciding on a sequence of operations to carry out to deal with a particular situation[43]. When people plan a course of action, they establish a programme of operations linked to when they will be done. Planning entails the operator reviewing a situation, judging what must be achieved taking account of the constraints on which plans will be permitted, then setting up a course of action allied to a course of monitoring to determine how intermediate states are being achieved.

Information requirements for diagnosis and planning

It is inappropriate to be too prescriptive about the information that needs to be provided to support *diagnosis* and *planning* because different people may use information in different ways. Most systems that malfunction provide the operator with different ways to carry out diagnosis. Symptoms manifest themselves in different ways and several different parameters that may be observed by the operator to reduce uncertainty may correlate with one another, so the same information can be gained by different routes. This means that different operators can use different strategies and be equally effective in reaching an appropriate decision. It also means that an individual operator could vary strategy from occasion to occasions depending on circumstances. For example, time pressure may warrant a different approach to situations that are less stressed; recent observations may prompt a different strategy for obtaining information than that which is usually followed in other circumstances.

It is obvious that if an operator wishes to follow one method of reasoning about a problem, yet the interface does not contain the required information or does not make such information easy to obtain, that operator will be at a disadvantage in reaching a decision in comparison with an operator whose strategy is supported by the interface. While it is possible to suggest information sets for the other general operational categories - action, communication and monitoring - diagnosis is far from straightforward.

One strategy for developing information requirements for *diagnostic* tasks is for the analyst to work out a set of preferred methods for dealing with common problems. For example, if a diagnostic strategy can be shown to entail first distinguishing between different parts of the system in which the problem might reside, then checking the status of certain items of equipment within suspect areas of the system, then an HTA of this strategy can be produced, which consists of a sequence of reading, monitoring and testing operations, which comply with the operational types listed in Table 9.1. In other words, to assess information requirements for diagnostic tasks, HTA can be used to set out how an expert examines the problem in terms of a step-by-step record of obtaining different sorts of information, including those actions necessary to obtain information not readily displayed. Then training can address the problem of training people to make decisions in a way which is consistent with these identified strategies.

Planning is, perhaps, even more problematic, because it can entail an element of creativity. Planning in wholly novel contexts is not something that can be tackled by HTA. However, more routine planning tasks, such as how to plan a return to normal service in a transportation system, how to recover target operating conditions in an industrial plant or how to plan a new office system can provide the analyst with precedents. Different instances of how such tasks were done in the past can provide the analyst with scenarios to analyse. These scenarios will enable tasks to be analysed, again revealing the action, communication and monitoring elements that were used. This will provide the analyst with evidence of the sort of information that was used to plan on previous occasions and also point to useful strategies that can be trained to help operators carry out these planning tasks in the future.

These suggestions for diagnosis and planning are not ideal, because they do not deal exhaustively with the range of possible solutions that operators could bring to problems. However, in view of the variations and complexities of diagnosis and planning, they provide a useful way forward in checking whether the information that people wish to use to make decisions is, indeed, provided. It also shows the kinds of strategy that they might use.

Identifying information requirements in plans

We can also obtain guidance to how operators use information to carry out their tasks by exploring the plans in HTA. Plans state the conditions when subordinate operations are carried out. In fixed sequence plans, the cue for each successive operation is the feedback that confirms that the previous operation has been completed. This means that *fixed sequence* plans provide no further information beyond the information that has already been identified by considering the operations - in accordance with Table 9.1. In *contingent fixed sequence* plans, a subsequent operation is carried out when a particular condition, such as an elapsed time or a target temperature shows that the operation is now due. In *choice* plans, information will be specified to indicate which options are followed. Thus, plans contain a number of cues and conditions that show the information designer various additional parameters upon which the operator is reliant.

Representing and situating information and control

Having adequate information available to an operator may be insufficient if the information is difficult to locate or difficult for the operator to interpret. Representation of information and control is a central issue in human factors. Ergonomics research deals with how instrumentation is represented. Environmental issues which how instruments and other signals are seen, heard, felt or smelled. Work organisation can affect task workload and the distribution of attention. Equally, human-computer interaction research has recognised the importance of task representation, taking advantage of the opportunities to program interfaces with greater flexibility to match the requirements of the task to the style and capabilities of human information processing. Thus, where older display technologies were confined to representing information in a limited manner, through digits, dials and recorders, computer technology provides many more design opportunities including representing system tools and information using *icons* and *metaphors*. Such ideas guide the design of a product to enable the user to anticipate what functions are available and how these are used. For example, *Macromedia Director™* , which is used to create animation, uses the metaphor of making a film to organise its different functions. Thus, the elements which are included in the program are called *cast members* and are organised in *casts*; the time-slots for animation are called *frames*; the action is assembled on the *stage;* and files are called *movies*. Ergonomics and HCI research have both been influential in helping to understand how best to design controls - how displays and controls may be designed to conform to the muscular, physiological and anatomical constraints of effort, and how they should facilitate psychological processes used in the task.[44]

An HTA carried out with a view to identifying information requirements may also supply other information about the task that the designer can use in deciding how information and controls should be represented. The HTA will provide the designer with information about task functions which can aid the choice in how information and control devices should be represented in order to help the operator recognise how these

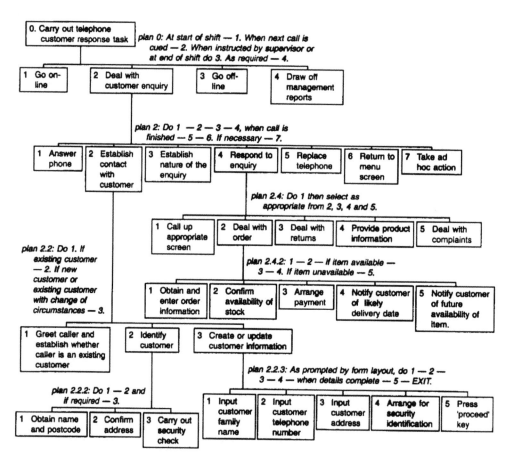

Figure 9.6   HTA of a tele-sales task in a call centre.

devices should be used. The plans also provide guidance about how information should be grouped as part of this representation. In menu systems, for example, the information required to support a particular function or activity should be presented on a single screen or, at least, screens which can be reached quickly from each other. The task hierarchy can guide which aspects of the task relate to each other in order to inform the locating of information on screens and specifying how screens need to be linked to enable the operator to progress the task satisfactorily.

These issues of interface design and navigation are beyond the scope of this book. However, the sorts of benefit for interface design that can be gained from using HTA are illustrated by reference to a small example. Figure 9.6 shows the HTA of a typical tele-sales task, where customers telephone a call-centre with their enquiries. They may be existing or new customers and their enquiries may range from making purchases to making complaints. The analysis shows general tasks of dealing with customers and dealing with management reports. In dealing with customers, the operator must identify the customers then identify the service that the customer requires. Figure 9.7 shows a possible configuration for a menu system to support this task. The top screen 1 offers a choice of dealing with a customer or calling off management reports. When using screen 2 to deal with the customer the operator is prompted to establish the caller's identity, moving to screen 4 to identify an existing customer or 5 to create a new customer file. When a customer identity is established, the operator establishes the purpose of the call and proceeds as appropriate with screens 7 to 11. For this sort of system it is also important to establish 'hot links' to enable the operator to move quickly between the main choice screens (i.e. 1, 2 and 6). This is a simplified example, but it shows a relationship between the HTA and a possible configuration of a computer menu system to support the task.

Concluding remarks

The importance of information and control is fundamental to the execution of tasks. Where the information and control facilities within a task are unsatisfactory, it can be the cause of serious problems. These include problems where there is insufficient information for an operator to control actions and where there is inadequate information for the operator to make decisions.

HTA offers opportunities for an analyst systematically to identify the information requirements in a task. Different types of operation need different sorts of information to support performance, including information to provide cues to act, identify instruments and controls, specify targets and provide feedback to the operator. Different types of actions, communications, monitoring and decision-making are identified, each type requiring different information sets to support performance. In addition, plans specify information concerned with making choices. When an HTA is complete, the analyst may use these categories to establish the set of information necessary to support performance.

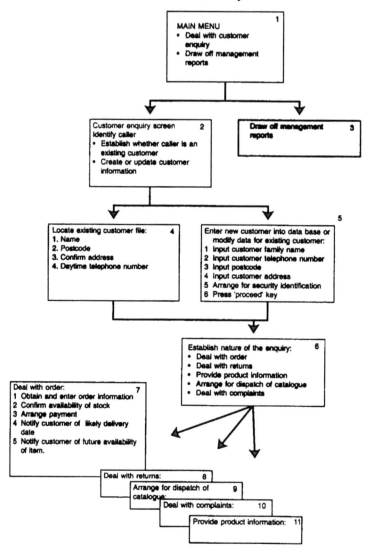

Figure 9.7   A possible way of organising menus for the tele-sales task in Figure 9.6. Each screen would also contain links to return the operator to other menu screens.

Providing sufficient information is not sufficient to guarantee effective performance. Task design must also ensure that information can be located, discriminated and interpreted. These issues have been central to interface design research. An example is given to show how some of the properties of HTA lend themselves to supporting these issues in interface design.

Information and controls are only useful if the operator chooses to use them. This depends on operator strategy. Strategies are influenced by the training the operator receives. Training is considered next in Chapter 10.

Chapter 10

HTA and training

Task analysis is vital in the proper development of training. First, it ensures that the aims of a training programme are focused on the operational requirements of the task for which it is being devised. It then provides the trainer with insights on how the task can be trained.

HTA is used to describe the task and then to explore the skills and knowledge that can help someone learn. It helps the training designer identify parts of the task to train and shows how these parts can be sequenced within a training programme.

Introduction[45]

Instruction and training can be regarded as the manipulation of conditions to enable learning. We often refer to the processes of working directly with the learner (or trainee) to facilitate learning as 'giving instruction', while the word 'training' refers to a set of organisational processes. These include analysis of needs, planning and developing conditions for learning, assessment and accreditation, and organising the delivery of instruction.

There are a number of different instructional methods that are encountered in practical training situations - sitting in classrooms; learning in simulators; using computer-based training packages; learning on the job under the guidance of an on-job instructor, coach, or mentor; or by working alone, thereby gaining experience. There is no 'one best way' to instruct people; different tasks, different circumstances, different learner preferences and different training resources warrant different approaches. Generally, a combination of instructional methods is advised in order meet a particular training need. This ensures that different parts of the task are mastered and the trainee develops the confidence to carry out the job alone.

This chapter is concerned with how HTA may be applied to practical problems of training. It will first consider a model of how people learn practical skills, which will be used to highlight where training staff can assist the learning process. Then it will consider the ways in which HTA may be used to support these training processes.

Learning practical skills

We learn things all the time about the world. If there is something that we cannot do, we have to work out how to do it. This may be through a combination of trial and error, using cues from how tools and equipment are designed and represented, our knowledge about the task or similar tasks, or with help from someone else. If, through these methods, we can deal with the task successfully, then on a subsequent occasion we may be able to deal with similar circumstances far more quickly, reliably and confidently. For all practical purposes, this improvement in capability constitutes learning, however it may have been achieved.

This can be illustrated with familiar examples. Many people, when unpacking a newly purchased electrical device - a toaster, an electric toothbrush, a calculator, a video-recorder, a personal computer - do not bother to read instructions, but learn to use the device through a process of exploration. They know that electrical devices need electrical power - it is assumed that plugging in the device to a power socket will provide this power. Switches labelled 'on/off', or switches that look like they will turn the device on or off, are usually good starting points. If a light comes on, or something starts whirring, then this provides feedback that the action was appropriate. If no such feedback is obtained, then the person will switch on the mains socket or seek another method for turning the device on. If none of these seem to work, the user will reluctantly look for the instruction booklet. Once the method of switching the device on and off has been discovered, the device can be switched on and off with confidence. It can now be assumed that the user has learned something.

When it has been switched on, using the device requires more selectivity. The user starts looking for facilities that are expected in such a device. In a tape recorder, the user will look for common functions of forward, reverse, rewind, search and record. These might be identified by written labels or symbols which suggest their usage. As various controls are tried, the user obtains feedback concerning whether the 'guess' was correct. If the guess was wrong, then the user hopes that no damage was done and looks for something else to try - or gives up and looks in the instruction manual. On a subsequent occasion when the user tries to do the same thing, a previous successful method may be remembered, and tried again. In this way, the action may become reinforced. If there was a substantial delay between first achieving such a goal and then subsequently attempting it, the user may forget what was successful first time round and need to explore the possibilities all over again. Confronted with a new challenge a person responds with a known solution or develops a new one.

This basic process of engaging in a task is set out as an HTA in Figure 10.1. The operator first looks for an existing way of responding. The operator who has already mastered the task can respond confidently with a proven method. If the operator is successful, attention will be directed elsewhere. If the operator is unsuccessful, then a new method must be sought. Evaluating the outcome will indicate if system feedback is unsatisfactory. Where a tried and tested method cannot be found or has been proven unsuitable, then the operator must develop a new method. This is now a problem solving activity where the operator reviews the facilities available and then plans a suitable response. The outcome is evaluated. If system feedback shows the method to be failing,

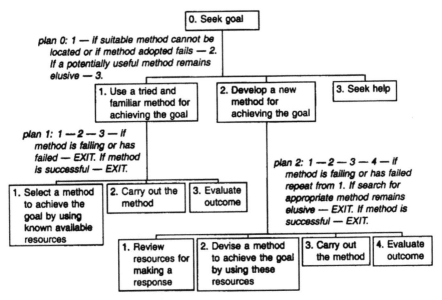

Figure 10.1   An HTA of learning by experience.

then a new method must be sought. If the operator despairs at being successful, then he or she will abandon the project or seek help[46]. This is not, strictly, a model of learning but a model of *carrying out operations*. Learners are not assumed to know what to do in a set of circumstances. When they do work out what to do and start to perform with more fluency, they are judged to be competent and, eventually, experts. This is a useful model to adopt when considering practical training, because learning depends upon what the individual learner does. If conditions are favourable, the learner will learn.

From this view of learning, it is a short step to see how *training* and *instruction* can help - learning can be facilitated by promoting conditions that facilitate learning, rather than hindering it. For example, if the learner is told which knowledge and principles could be helpful in working out how to achieve a goal, then he or she will be more able to achieve success. Equally, if the learner is given a second opportunity to attempt the goal, then he or she will be able to correct previous errors or consolidate successful responses.

Much training research has focused on prescribing how different sorts of training should be done and the relative merits of different training methods. Practical considerations rarely enable straightforward prescriptions to be followed. Different training methods have different strengths and weaknesses according to the context in which they are applied. To develop training, the trainer or instructor must to try to understand what is happening to a trainee who is trying to learn, rather than pursue a dogmatic set of training methods. In this way, training design is concerned with *managing* the learning process, rather than following a particular dogmatic approach.

An informal training intervention

To illustrate the practical aspects of learning and instruction, we shall consider a simple and generally familiar occupational task, that of working in a shop. A trainee shop assistant must learn about performing transactions with customers, tidying stock, cleaning the shop, filing orders and so on. Some of these things can be dealt with together - they are responses to different sets of events and some parts of the task are probably too complicated for an inexperienced person to understand all at once.

The task analysis

Figure 10.2 shows a typical, though simplified, HTA of a shop assistant's task. Some parts are concerned with selling to customers, but there are also many parts concerned with good 'housekeeping'. Carrying out housekeeping routines such as cleaning the shop and keeping shelves tidy are things that are done when there are no customers to serve. Customers are served when they present themselves in the shop and request service. Some elements of the task are critical and, even unrecoverable. Making a mistake during a payment transaction could lose money, whereas placing an item on a wrong shelf can be rectified.

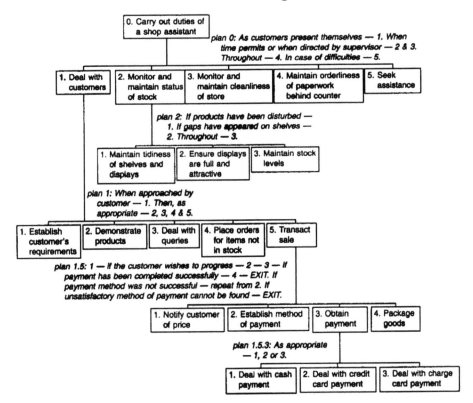

Figure 10.2 HTA of a shop assistant's task.

Instructional methods

Some ideas for training these various parts of the task are set out in the HTA table in Table 10.1. If the new trainee is a young person with little experience of working in organised systems or even of handling different ways of paying for things, then he or she would need extra tuition concerning, for example, how to distinguish between credit cards and charge cards and how each must be dealt with. A more experienced person may already be competent. If they have worked in shops before, they may even have experience of the stock control system used, at least they may have some concepts of stock control which can be used to explain the new system. There are several instances in this task where one would wish to give some off-job training. For example, concepts of different sorts of payment card and the shops stock control system can be explained.

So far, the HTA has helped in two main respects. First it has provided an account of the task to be trained. This ensures that training is directed towards to real needs of the task. Thus, important things will not be left out and irrelevant things need not be included. Second, it enables the training analyst to scrutinise the task systematically to determine how each part could be trained. A third issue, which we will now consider, is how the training should be sequenced.

Table 10.1   The HTA table for the shopping task in Figure 10.2.

Task Analysis		Training suggestions
0 Carry out duties of a shop assistant plan 0: When approached by customers — 1. Throughout or when time is available — 2, 3 & 4. If you are uncertain of actions — 5.		
1 Deal with customers	yes	Courtesy in all customer transactions is stressed.
2 Monitor and maintain status of stock	yes	
3 Monitor and maintain cleanliness of store	no	Diligence is emphasised and a good example is
4 Maintain orderliness of paperwork behind counter	no	maintained by all staff.
5 Seek assistance	no	Trainee should ask advice whenever uncertain. Assistance should always be provided in a way to promote the trainee's learning.
1 Deal with customers plan 1: When approached by customer — 1. Then, as appropriate — 2, 3, 4 & 5.		
1 Establish customer's requirements	no	Trainee is required to observe experienced staff in operation. Early close supervision is given on-job.
2 Demonstrate products	no	Opportunity given to enable trainee to gain familiarity with products.
3 Deal with queries	no	Product sheets to be studied.
4 Place orders for items not in stock	no	Procedures demonstrated off-job where practice is also given. Supervised on-job.
5 Transact sale	yes	
2 Monitor and maintain status of stock plan 2: If products have been disturbed — 1. If gaps have appeared on shelves — 2. Throughout — 3.		
1 Maintain tidiness of shelves and displays	no	Diligence is emphasised and a good example is
2 Ensure displays are full and attractive	no	maintained by all staff.
3 Maintain stock levels	no	Stock control system is explained off-job. On-job task is demonstrated and supervised closely.
1.5 Transact sale plan 1.5: 1 — if the customer wishes to progress — 2 — 3 — if payment has been completed successfully — 4 — EXIT. If payment method was not successful — repeat from 2. If unsatisfactory method of payment cannot be found — EXIT.	no	
1 Notify customer of price	no	Courtesy in all customer transactions is stressed
2 Establish method of payment	no	
3 Obtain payment	yes	
4 Package goods	no	Demonstration and practice in stock room. Supervision early in work experience.
1.5. Obtain payment plan 1.5.3: As appropriate — 1, 2 or 3.		
1 Deal with cash payment	no	Obtaining payment is a critical task and requires that
2 Deal with credit card payment	no	terminology and systems are properly understood.
3 Deal with charge card payment	no	Procedures can be practised off-job.

Sequencing training

A good start to any training programme is to provide the trainee with an understanding of the kinds of things the job will include. The top level of the HTA provides a useful task organiser or an outline of a job description, which may help the trainee see how various parts of the task fit into context. There is no obvious guide to where to concentrate next. Strictly speaking, it does not matter whether the trainee turns attention to stock maintenance issues or to dealing with customers. However, by carefully considering each of these, preferences may emerge.

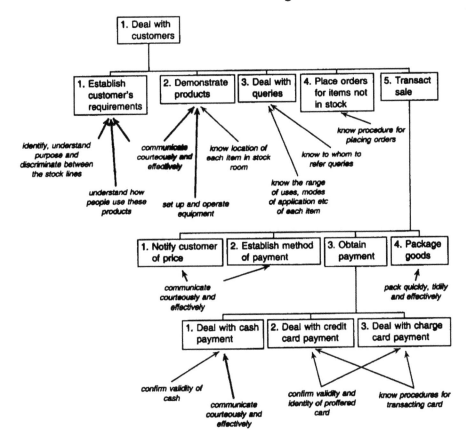

Figure 10.3   Skills and knowledge mapped onto the task analysis.

The sequence of training need be constrained only if there is reason to believe that practising some things first will make other things easier to learn. One basis for constraint is ensuring that appropriate concepts are understood before crucial parts of the task are practised. Figure 10.3 shows how knowledge about the task maps on to the different things that the shop assistant has to do.

For example, to 'demonstrate products' (operation 2) entails the trainee knowing how to communicate courteously and effectively, set up and operate equipment and know the location of each item of stock. It is possible that these three things could be learned as part of the process of demonstrating products, although this could lead to a number of errors to be corrected by the instructor in front of the customer. It would make more sense for the trainee to master some of this material before practising the overall task. Equally 'obtaining payment' relies on the trainee knowing how to deal with each of three different sorts of payment - getting this right is critical for the profitability of the shop. There is particular reason, therefore, for this to be done well, so there is reason to teach these component skills beforehand.

The trainer must decide whether a trainee will learn to carry out a particular operation all at once or first be taught some of the component skills in order to make practising the full task easier. Different contexts may warrant different strategies. The trainer would consider:

- the operational consequences of trying to learn these things whilst practising the task - primarily, risk to people and equipment;

- the suitability of the operational environment for enabling practice and instruction - control over events, task feedback, recovery from error;

- the personal consequences for a trainee in exposing him or her to the risk of public failure - embarrassment, stress, reduced motivation;

- the disadvantage for the trainee in trying to master too much at one time - this may vary between individuals as some people know more than others and some people are more able than others to cope with uncertainties in the learning situation.

A further factor that may influence the choice of what the trainee should do first concerns commercial demands. In training shop skills it may be preferred that the new starter first deals with stock control and housekeeping tasks in order to learn something about the product. However, there will be occasions, when other staff are unavailable when the new starter is required to deal with customers. Shop management must assess the risks entailed with this. Errors committed in shop work are often recoverable and there is time for the shop worker to seek assistance in cases of difficulty. Other contexts - medical, transportation and some areas of production - where errors are not so easy to recover and help is not always to hand - permit less flexibility in how people are trained. Contexts such as these are characterised by longer periods of training and by a greater willingness to invest in training resources.

Transfer within the task

One economy in training may be gained by examining how different parts of the task share common skills and knowledge. From Figure 10.3 it is clear that some elements of knowledge and skill are specific to particular operations, while others, such as 'dealing courteously with customers' are common to all customer transactions. One obvious example is 'communicating effectively'. This means that if communication is mastered and demonstrated in any of these particular contexts, the trainee may be able to apply these skills elsewhere.

Assessment

An important constituent of any training programme is the assessment of the trainee's competence. Any practical assessment of competency at a task must entail judgements about how well operators perform in the real world. In the shop assistant case, a trainee may never encounter certain methods of payment during the period when he or she is being assessed in the shop. We may form a judgement that the trainee is able to follow the correct procedure for each of the payment methods by observing performance in the peace and quite of a training room. However, when confronting a real customer, the trainee might panic. Therefore, a combination of assessment methods must be utilised. It makes sense, first to establish that the trainee has mastered the procedures for each of the payment methods. Then it is important to establish that, when encountering a customer in the shop, the trainee is able to select the appropriate method and carry it out without error. Provided this 'live' assessment is done a few times, it can be reasonably inferred that the trainee is both competent in remembering the requisite methods and confident in working in an unflustered manner. On such evidence, the assessor judges whether the trainee is competent in this operation.

The same applies in fault diagnosis in industrial plant, medical diagnosis and air-traffic control. In each of these situations, the events that any trainee will experience on the job - during instruction and assessment - will be limited and will not provide sufficient evidence to make wider inferences about competence. To become confident in a trainee's ability, an assessor needs evidence of breadth of knowledge and skill, perhaps obtained from a range of tests under simulated conditions and also evidence of the trainee's. confidence in the real situation.

The discussion so far has dealt with training in a flexible way. Often training has to be more formal, because tasks are more complex and ensuring competence is more critical. Training conditions have to be devised carefully because there is a need to prepare training material, including devising the environments in which people can practise. Training programmes must be planned and training materials devised or procured according to budget and in sufficient time. They also may need to be auditable such that their benefits can be demonstrated to managers within the organisation and outside inspectors.

Elements of a formal training programme

The shop assistant's task provided an informal example of training. Even so, it contained many of the elements that are encountered in more complex training programmes. Practical training literature is abound with methodologies prescribing what trainers should do to develop training. Training practices should be tailored to match the requirements of the organisation for which the training is being developed. To illustrate the potential contribution of HTA, the principal elements of a training design and development process will be described.

Even formal training programmes should focus on the experiences given to trainees to help them learn, rather than following a particular set of prescriptions. Figure 10.4 shows this process of instruction by setting out a range of instructional activities in conjunction with the trainee's learning experience. The activities in the grey panel set out what needs to be done to provide the direct support for the trainee practising the task. These include preparing the trainee, setting the trainee to practise and providing feedback once the session is complete. The activities outside of the grey panel show a number of things the instructor must do to prepare and record instruction. A full training programme will repeat this cycle for different training elements until the overall task is

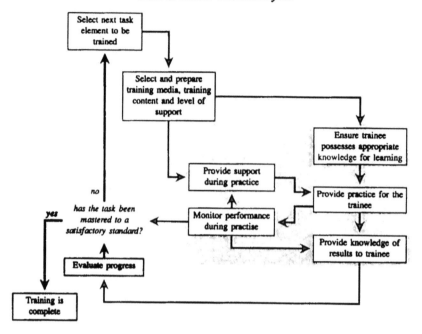

Figure 10.4   Interaction between the instructional cycle and the learning process.

mastered. Developing a training programme entails identifying what needs to be trained, who is to be trained, how parts of the task should be related to each other and what methods should be used to deliver instruction.

Stages in developing a training programme

Table 10.2 lists the main stages that must be considered in the design and development of a training programme.[47]

1.   Assess training needs

'Training needs analysis' is a common phrase in the training world, but one which can mean different things. 'Training need' can be used to refer to whether training is necessary to support a system and to identifying the sorts of training that need to be provided. It can also refer to the judgement concerning who is to be trained to support a system adequately. This will reflect numbers of personnel and specify their grading. This is important in justifying training resources. Thus, where many people are to be trained, expenditure on expensive training solutions will be easier to justify that where few people are to be trained. Even where few people are to be trained, expense may be justified if parts of the task are critical. Part of training needs analysis, therefore entails making some sort of training inventory to establish manning levels.

HTA obviously contributes to understanding what a task involves and, since this is done in a systematic way, the analyst can identify alternatives to training. A serious

Table 10.2   *Typical stages in a training design and development process.*

```
1  Assess training needs
    1.1. Identify numbers of staff to be trained
    1.2. Establish existing capabilities of staff to be trained
    1.3. Analyse tasks, skills and knowledge to be trained
2  Design training programme
    2.1. Specify training criteria for task elements
        2.1.1. To establish reliable invariant performance
        2.1.2. To establish flexible performance
        2.1.3. To develop confidence
    2.2. Set out the parts of tasks and the training programme
    2.3. Specify training sessions
        2.3.1. Specify start criteria
        2.3.2. Develop pre-instructional material
            2.3.2.1. Classroom learning in groups
            2.3.2.2. Computer-based learning
            2.3.2.3. Self-learning manuals
        2.3.3. Providing conditions for practice
            2.3.3.1. On-job training
                2.3.3.1.1. On-job instruction
                2.3.3.1.2. Experience/development
            2.3.3.2. Simulation
                2.3.3.2.1. Part-task trainers
                2.3.3.2.2. Full mission simulator
        2.3.4. Establish training criteria
    2.4. Identify training methods and media
    2.5. Review training options against resources
3  Develop training programme
    3.1. Develop training staff
    3.2. Develop training materials
    3.3. Pilot training materials/programme
    3.4. Review and revise programme
4  Deliver training
5  Monitor continued effectiveness of training
    5.1. Review performance of staff
    5.2. Review changes in technology, legislation and procedures
    5.3. Review adequacy of current training provision
```

training project, concerned with providing competent staff to work within a system should also look for non-training alternatives either to reduce the amount of training needed or to ensure that any tasks to be trained are well-designed and can be more easily learned.

2. Design training programme

An important early step in any training design procedure, is to understand the criteria of the skill to be developed. In some cases, it is a requirement that reliable invariant performance is developed, while in other situations it is required that the operator demonstrates some versatility. Versatility enables the operator to vary performance in accordance with circumstances. The HTA will enable the analyst to judge how different parts of the task should be treated in this respect. This can be recorded in the analysis table.

All training programmes entail identifying different parts of the task to be trained separately and then building up competence of the task overall. The structure of HTA is particularly suitable for helping the training designer decide how to organise the training programme in this respect. The issues of part-task training will be developed separately shortly. For each part to be trained, the training designer must specify the readiness of the trainee for the training to be undertaken, the material and methods to be used to provide the trainee with appropriate task knowledge and skill for the learning to come, the manner in which the task element will be practised and the performance criteria against which the trainee will be judged.

Knowledge and constituent skills can be taught to the trainee face-to-face in a classroom or with an on-job instructor, or using some recorded media, such as computer-based-learning, an instructional video or a written guide. The decision to choose each of these alternatives will depend, to some extent, on the resources available and the extent to which management is willing to invest further in training resources.

Critical to learning occupational tasks is practising the task. Practice, as has been emphasised earlier in this chapter, enables the trainee to apply his or her knowledge and skills to resolve a new situation and then consolidate the lessons learned. Practice also demonstrates to the trainer the extent to which the trainee has mastered the task. The training designer must decide how best to provide opportunities for practise. Tasks can be practised on the job or in an artificial environment, i.e. through simulation.

When the training designer has decided which methods and media are most suited to dealing with each of the operations identified, the list should be reviewed to make final decisions concerning how to proceed. It may be inappropriate to recommend training options which entail using resources which can only be procured at a cost. On the other hand, if equipment must be procured to ensure that some parts of the task are trained to a satisfactory standard then finding other uses for this technology will not add significant cost to the project.

3.   Develop and Deliver training programme

When the design for a training programme has been judged to be satisfactory, it must be developed in a form to be delivered. This is a project management task. It entails recruiting and training sufficient staff to develop training materials and deliver -the training. The training materials must be developed to time. Then the materials need to be tried out and revised as necessary before going into full use. Piloting training materials is, generally, essential to ensure that they can be used properly and that they can achieve required training benefits. If there are any problems, this is an opportunity to make any modifications necessary.

4. & 5.   Deliver and monitor continued effectiveness of training

When a satisfactory training programme has been developed, it can then be put to use. Whenever a training programme is in use, there should also be a system for continually monitoring its adequacy. No matter how good was the original set of training materials, circumstances change. There can be changes in the types of people being

Table 10.3   Summary of the main benefits of HTA in training design.

Training design decision	Contribution from HTA
1 Assess training needs	• Examining tasks in their operating context.
	• Setting out capabilities required to enable comparison with existing skills.
	• Identifying suitable training media and methods.
2 Design training programme	• Identifying training criteria for each part of the task.
	• Identifying parts of the task to train in a part-task training programme.
	• Providing a framework for specifying how each task part can be trained.
	• Scheduling the training of parts to enable overall task mastery.
	• Identifying information and controls to include in simulators and part-task simulators.

recruited. There can be changes in production methods, in the procedures adopted by staff, in the equipment used in the task, demands of the market which affect performance requirement and any legislation governing the context in which people work. It is also important to review developments in training technology since these can bring improvements to training provision.

Of these five main steps, the benefits of HTA for training focuses primarily on the training design phases - stages 1 and 2. These establish training content so that training effort is properly justified. Then the training designer can focus systematically to decide how each part of the task should be trained and which media should be used. There is both economy and consistency if the trainer can move through each stage of the training using the same basic representation of the task. In this way, the task description established at the outset with the client can be used to ensure that training decisions remain consistent with operational needs. These benefits are summarised in Table 10.3.

Two issues warrant further examination. These are *part-task training* and *simulation training*.

Part-task training[48]

'Part-task training' is the term given to the process of breaking a task to be mastered into separate chunks, so that the trainee can focus attention onto parts of the task and then progress to mastering the overall task. Part-task training is undertaken because most real tasks are too complex or too diverse to be mastered at one time. Virtually all practical training entails part-task training to some extent. Hence, trainers develop syllabuses or learning programmes where the trainee commences from fundamentals and then works through material of increasing complexity and diversity. It is important

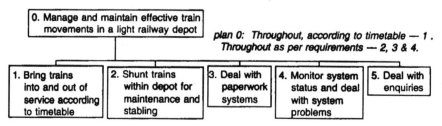

Figure 10.5   Top level in the HTA of managing trains in a light rail depot.

to understand how 'parts' of tasks can be identified to be taught separately in a way that will transfer to the full task.

Two different issues prompt the need for part-task training. The first is that in some jobs only certain events can occur at any particular time, so only those parts of the task can be practised at any one time. For example, when the time comes for a customer to pay the shop assistant only one method of payment is likely to be pursued at any one time, so only one method can be practised. A process plant operator will sometimes be concerned with starting a plant up and on other occasions will be monitoring plant parameters. A nurse will sometimes be involved in managing a treatment and sometimes with making a test, but not at the same time. These things occur at different phases of the work and will be acquired separately. Figure 10.5 shows the top level of the HTA of managing trains in a light rail depot, where trains are parked, washed and maintained when they are not on the main track. Plan 0 indicates the different operations that are carried out on different occasions when different events arise. Contrast this with the simple procedure for changing the cartridge on a printer, where, once started each step is carried through in sequence without variation.

The second justification for part-task training arises as a strategy for helping the trainee to master things that would, otherwise, be too complex to master all at once. The monitoring operation in Figure 10.6 is all done at the same time, but there is a lot for the new starter to master. In each case, the operator must know what to look at, for example, to 'Monitor location of all personnel and deal with transgressions' (operation

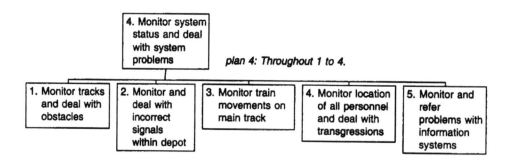

Figure 10.6   Monitoring operations within the depot.

4) entails knowing how and where to look - through different windows, into different corners of the depot and also using the video camera showing different parts of the depot. The operator must also know what constitutes a transgression and how different transgressions must be dealt with. All of these entail rules and strategies that the operator must learn at the outset, then practise applying. Each of the monitoring operations in Figure 10.6 entail similar considerations.

Progressing part-task training

How to progress in a part-task training programme is of prime importance. One way in this example is for the rail depot trainee to be allowed to deal with each monitoring activity in turn, for example, concentrate on monitoring train movements. The trainee can, thus, concentrate on learning the rules and strategies for that operation and then progress when confident and competent. Then, when each operation is mastered, the instructor can tell the trainee to deal with all of them together. Unfortunately, this may not work in view of the nature of monitoring tasks. Monitoring several subsystems at once is a complex skill which is more than just the sum of the individual skills. When practised separately, the trainee will focus exclusively on a particular subsystem. When practised together, the operator must divide attention between subsystems. The strategy that the trainee develops to deal with *each* monitoring activity separately may be excessive when having to time-share with other monitoring duties. The real monitoring skill required is that the operator can infer the status of the subsystem with a sufficient, but not excessive, amount of information. Moreover, there is a strategic element in distributing attention between several subsystems. The operator who is monitoring obstacles on the tracks and sees that they are clear and that there is no current activity likely to cause obstacles infers that obstacles are unlikely in the near future. If the same operator sees some staff close to part of the track, then greater attention must be paid to this aspect until the risk is reduced. The operator can, for the time being, give priority to monitoring the movement of personnel at the expense of worrying about obstacles on the track.

The same considerations apply in all contexts. A nurse on a ward must keep an eye on each of the patients in his or her charge. If one patient shows signs of concern, then greater attention may be directed towards that patient, but this cannot be done at the expense of the others. Swimming pool attendants may maintain a special watch on a group of particularly boisterous children, but they must also maintain adequate surveillance of other people using the pool. The air-traffic controller will need to focus attention on potential conflicts, but must also be alert to other features to monitor. Thus monitoring several systems is not simply a matter of aggregating different monitoring skills practised separately. The separate skills must be suited to the overall demands of the task and the operator must deploy these skills strategically.

There are different ways of progressing part-task training which have received attention in training research. These different methods are simply possibilities for arranging a progression of training. We should not assume that some are better than others for all tasks. Figure 10.7 shows three typical part-task training arrangements that could, in principle be applied to training the five monitoring operations shown in Figure 10.6. In *pure-part* training each monitoring task would be practised separately. Here

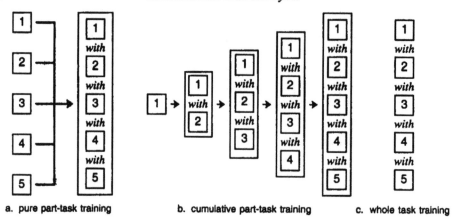

a. pure part-task training   b. cumulative part-task training   c. whole task training

Figure10.7   Some part-task training arrangements.

there is the risk that strategies that the operator adopts become too complex when the task is practised as a whole and that the operator does not learn to vary strategy according to risk. In whole task training, where the trainee is required to practice everything at the outset there may be too many things for the trainee to deal with at once and he or she may learn very little, or learn at a very slow rate. *Cumulative* part-task training may provide a useful compromise. The trainee practises one monitoring skill then, before long, adds another an so on. In this way the trainee is learning to time share. This might provide the best of both worlds. There are other possibilities. For example, in the cumulative part-task training option, the trainee could progress by adding two monitoring tasks at a time. Another option is to introduce each new monitoring task with a short period of practice of that task on its own, quickly followed by practising all of the accumulated tasks so far.

It is not appropriate to be dogmatic concerning which method is most effective for a particular task. The training designer must carefully consider the nature of the task and how the parts relate to each other. Ideally, the training designer would test and fine tune the chosen method.

HTA and part-task training

The role of HTA in this design activity is concerned with deciding which parts of the task are most useful for the training programme and in shedding some light on how these parts relate to one another. For example, in Figure 10.6, the training designer can see that five monitoring operations are carried out together. The training designer should also be aware, from Figure 10.5, that the monitoring task itself must be carried out while other things are being done. This time sharing is a common requirement. If a baby in neonatal intensive care is in crisis and the main effort is devoted to dealing with that crisis, it is wrong to neglect other monitoring activities because other problems for the baby can arise. Equally, if an air-traffic controller recognises a potential conflict between three aircraft in a sector, resolving the conflict must be time-shared with managing the remaining air space. If nurses or air-traffic controllers or railway staff develop monitoring strategies which require too much of their attention such that other tasks are neglected or monitoring suffers when other things are done, then they are not doing their jobs properly.

Suboperations, such as the monitoring operations so far discussed, are useful parts of tasks. Another useful way of partitioning the overall task is considering different types of event and controlling the rate of occurrence of monitoring problems to be detected. If it were possible to arrange matters such that the only problems that could arise during a period of instruction in the railway task were concerned with the signalling system, then the trainee could be set to practise the whole task but would, in the event, only deal with a subset of events. This means that the trainee could consolidate these skills before the instructor increased the possible set of events. Equally, if the rate of occurrence of events could be managed in accordance with how well a trainee could cope, then a new trainee would experience a benign environment and gradually deal with an increasingly challenging world. By inspecting the plans in HTA, the analyst can identify different sorts of events affecting the current demands presented to a trainee.

In most real tasks it is impossible to organise the world so conveniently, but where tasks are *simulated*, this is perfectly possible. Consider the supermarket checkout task shown in Figure 3.5. A new checkout operator could be taught using a whole task training method, in the sense of practising the whole cycle, but could be helped by the careful introduction of events to deal with. A first customer could have a small order comprising straightforward bar-coded items and then pay with cash. As the trainee progresses, orders could get more varied, breakages could arise, payment methods could vary and 'customers' (that is, people role-playing customers) could vary in how truculent they were.

Part-task training is a useful practical method for dealing with complex tasks and HTA provides the training designer with a number of ways in which tasks can be taught by parts which build to mastery of the whole task.

Simulation for training[49]

Simulation for training is the representation of features of a real situation to enable practice of a task. Training of all tasks requires some opportunity for practice. Simulation is employed in training or assessment of personnel when practice in the real situation is risky to life, health, plant and product or when satisfactory practice of certain situations cannot be provided, for example, for unlikely yet critical events. Simulators are used to train airline pilots to enable them to practice in safety. They are also used, for the same reasons, to train power plant operators, police overseeing crowd control, ship's pilots, tank commanders and medical staff. Surgeons may use simulators to practise new surgical techniques.

Simulation is often used to provide a means by which infrequent, yet critical events can be presented to trainees for practice. For example, possible but unlikely scenarios can be offered to a nuclear power plant operator or an airline pilot, to provide practice in dealing with unlikely, yet critical events. Simulation may also be adopted when the real task cannot provide a suitable training environment. Thus, it may be impossible to simplify the task for the trainee during the early stages of instruction, or the system feedback provided in the real situation may be too delayed to help the trainee learn and so additional feedback must be provided by the instructor.

Manipulating fidelity in simulation

Much emphasis is placed on simulators being as faithful as possible to the situations they are representing, and simulator manufacturers place great store on this aspect when promoting their enhanced graphics capability, for example. It is often assumed that by practising tasks in an artificial environment, which represents the same information cues, control facilities, feedback and environment, the trainee will be able to apply the same skill in the real task. But if the fidelity of the simulator is reduced in any way, then the trainee may develop a skill in the simulator which cannot transfer to the real task. In simulating surgical tasks, in minimal access surgery or defibrillation for example, it is often difficult to be confident that the artificial cues created properly reflect the 'feel' of the human body responding or resisting. Therefore, if the trainee becomes highly skilled in the simulator, the skill may not transfer to the real world which responds differently. Indeed, all simulation used for training entails some aspects of reduced fidelity, if only in terms of the anxieties that people feel when dealing with the real situation.

In cases where complete fidelity cannot be achieved, the trainer might be advised to reduce fidelity further to remove reliance on developing a skill that will not transfer successfully. Sometimes it is helpful to simplify or clarify an information display to help the trainee focus on critical aspects of the task and to augment feedback to enable the trainee to see the consequences of action more clearly. In many situations, it is essential that the trainer is careful to help shape the trainee's strategies through provision of appropriate knowledge, control of the events practised and careful attention to the strategies the trainee is developing, rather than simply allowing the trainee to develop any strategy that appears to work in the simulator. Simulation of lower fidelity may still be sufficient to allow the trainee to master a skill or strategy that can be applied and developed in the real situation. Cognitive task analysis methods may need to be employed in situations where the training designer has to try to understand the underlying cognitive strategies that the operator might be using.

A further crucial aspect of fidelity concerns the schedule of events. If a simulator were designed to present events with the same frequency and likelihood as the real situation, then infrequent events in the real world would occur just as infrequently in the simulator. Often, the justification for simulation is to provide the trainee with experience of events that occur infrequently, to master critical skills that would, otherwise, never be experienced. This creates a major problem for the training designer. In practical tasks, people often gear their responses to expectation, so the trainee learns to anticipate events in the simulator but cannot transfer these expectations to the real world.

Designing simulation for training should never simply be a matter of trying to replicate a working environment. It should be part of a process of designing training.

There may be several parts of the task that can use the same *simulator,* but the simulator must satisfy the requirements for *each* of the parts for which it is designed.

Were a simulator to be devised to train the railway depot supervision task in Figure 10.5, it would be appropriate to replicate displays showing the timetable and which trains were in service, the movement of trains within the depot and displays concerned with signalling. We would wish to replicate the paperwork system and include a telephone through which simulated enquiries would be fed. It may be judged that providing a simulation of the depot, through which the trainee can monitor obstacles or personnel on the tracks, is too expensive and that this should be left to be practised on the job.

To practise operation 2, concerned with identifying incorrect signals, the trainee could be presented with a display showing good and bad signal patterns in order to practise distinguishing acceptable signals from unacceptable signal patterns. This could be a low fidelity simulation, carried out on a personal computer. Other monitoring tasks can be taught at this fundamental level in a similar way. As the basic procedures for conducting these different forms of monitoring start to be learned, the trainee needs to deal with a more complex environment in which other monitoring activities can be incorporated - so that the composite skills of monitoring can be learned, as described earlier in this chapter. The training arguments concerning monitoring, discussed earlier in this chapter, pointed out that the monitoring task entailed the trainee learning to coordinate all of the monitoring activities, otherwise the trainee would develop skills that cannot transfer to the real task because of workload. Therefore, even though monitoring obstacles and personnel will be practised in the real task, the simulation must still contain some token exercise to occupy the trainee in a 'quasi-monitoring' activity to prevent the trainee devoting too much time to making the other monitoring decisions. This quasi-monitoring activity may entail the trainee having to move to a position representing the control-room window, in order to make a judgement about a signal from the trainee.

By treating simulation design as a component of training design, the analyst will maintain a proper focus on the requirements of the trainee mastering a skill.

HTA and the design of simulation for training

The role of HTA in the development of simulation rests with the contribution that HTA can make to training design training design, as set out in Table 10.3 - identifying training needs, parts of tasks, prerequisite knowledge and skill, training methods. If has an additional important benefit in that the information and controls that have to be represented in the simulation can be identified by examining operations and plans, following the principles discussed in Chapter 9.

Concluding comment

HTA has always had a close relationship to training design, but to understand this, it is necessary to appreciate how people learn in practical situations. *Learning* is a process in which a person attempts to deal with a task that he or she may not have dealt with successfully before. In dealing with such a task, the person uses prior knowledge and skill and information provided via the interface to determine which control actions will be appropriate. Task feedback shows the extent to which the trainee has been successful. Carrying out the same task with greater success on a later occasion implies that learning has taken place. *Training* can be thought of as the manipulation of conditions to facilitate learning. Thus the trainer assists the learner in different ways, including ensuring adequate prior knowledge and skill, a suitable environment in which to practise, help in dealing with the task and feedback to evaluate and consolidate learning. Another important function of training is to ensure that events for practise are scheduled to aid the learning process and consolidate retention of skill.

These ideas were used to set out the main activities involved in training for which HTA can be shown to provide effective support. By using HTA as a basis for training design, the trainer can justify where training should be applied and break the task into meaningful 'parts' which can then be taught separately and developed into a general task competency. The HTA can also enable the trainer to focus on various parts of the task to identify the skills and knowledge to facilitate the acquisition of skill.

Chapter 11 will deal with an area related to training, namely the development of job-aids and other user support documentation. Often, providing an operator with a support document means that training effort can be reduced. Moreover, practical training can often progress by providing the trainee with a manual to guide performance during the early stages of experience.

Chapter 11

Designing support documentation

Support documentation such as procedural guides, decision aids, manuals and other job-aids, provide an important method for helping operational staff accomplish their goals. It is important for the analyst to make appropriate judgements about where such aids are suitable, and then develop the job-aid to meet the requirements of the task.

The role of HTA in providing a coherent approach to job-aid development is demonstrated. It shows how HTA can be used to identify the relationship between training and the job-aid and then design the job-aid itself.

Introduction[50]

Chapter 11 dealt with training as a means of helping the operator acquire skills necessary to carry out tasks. *Training* implies that operators will be helped to remember or work out what to do themselves in a given set of circumstances. Another way is to provide operators with a *support document, job-aid or help-facility* to guide them. The term 'support document' is not common parlance and so, 'job-aids' will generally be used in preference in thischapter to refer to all support documents.

There are a variety of types of job-aid used to support performance in different ways. *Manuals* are usually reference documents; *procedural guides* serve to provide the operator with step by step-guidance in carrying out a task; *decision-aids* help the operator to distinguish between a range of options in deciding what to do next; *checklists* remind the operator of several things that need to be done. There are also record systems in which the operator routinely records specified information or writes down whatever seems likely to be important for people on later shifts.

From an operational perspective, it does not matter whether an operator's response is established with or without assistance from using a job-aid, provided that goals are met to a satisfactory standard. However, there are cases when using a job-aid can disrupt performance and cases where the job-aid can ensure consistent and reliable performance. Therefore, one of the decisions that the designer or analyst must make concerns when the use of job-aids is recommended. The choice is rarely between training people or providing them with job-aids on their own, but between training on its own and training in a supported environment where the job-aid is provided. That is, in most cases some training must be given along with the job-aid - to show how the job-aid is to be used, to ensure that the content of the aid can be understood and to ensure that the operator can carry out the instructions that the job-aid prescribes.

The exceptions to this are generally concerned with providing instructions for public utilities or domestic and commercial devices that people will use intermittently. For example, public telephones, ticket machines, bank machines, video recorders, photocopiers and fire extinguishers are all devices that members of the public might be called on to use in specific circumstances but for which they would not expect to undertake formal instruction. Despite this, people often call upon a friend or colleague to provide such instruction. When designing for such situations, assumptions must still be made about how the user employs certain terms and what sorts of behaviour these terms will provoke. In other words, job-aids always rely on some form of skill being supplied by the person using them, whether or not this is explicitly trained.

In developing a job-aid, then, the designer must:

- identify where a job-aid could be helpful and what form it should take
- decide which part or parts of the task should be supported with the job-aid
- decide how the content of the job-aid should be expressed
- decide how the job-aid should be organised.

In addition to these steps, it should go without saying that the task being supported should be properly understood to ensure that the job-aid represents the procedures and decisions necessary to operate the system.

HTA can be helpful in supporting each of these decisions. Of particular significance is that job-aids are concerned with helping the operator decide what to do next. This relates closely to plans in HTA. Therefore, completing an HTA provides useful guidance concerning how job-aids can be presented and structured.

Common types of support documentation

Job-aids come in many different forms and serve different purposes. Invariably, though, they are concerned with helping an operator know what to do next. Therefore, they will relate to plans in HTA. Different sorts of support documentation will reflect the different sorts of plans in HTA.

Procedural guides

Procedural guides are provided to ensure that the user/operator carries out tasks in a set way. Apart from labels on equipment, which will not be dealt with here, procedural guides are the most common and straightforward of all job-aids. Procedural guides may be used where it is felt that deviation from certain sequences of action may compromise productivity or safety. Thus, procedural guides are provided to help people operate equipment which is only used intermittently, for example, photocopiers, or to support crucial tasks that are not practised frequently, for example certain maintenance tasks or emergency procedures.

The simplest form of procedure is where the user/operator carries out one action, then when it complete, carries out the next action and so on. Figure 11.1 shows the HTA of part of developing a black and white film. Figure 11.2 (a) shows a procedural guide to support the process of preparing the developer, translated directly from operation 3.1. More complex procedures require the operator to carry out one step, then monitor a specified parameter to await the conditions for the next step. Thus, Figure 11.2 (b) shows a job-aid to support operation 3. Plans in HTA contain much of the information needed to organise job-aids and the statements used to describe operations can often be transferred to the job-aid with little modification. Thus, the plan will specify the next action to take. The plan also indicates the cue to the next action, either feedback to signal that the previous action is complete or a separate cue to prompt the next action, in the case of contingent fixed sequence tasks.

Checklists

Checklists are provided to prompt the operator about what has to be done, usually in cases where actions are important but where their order of execution is not. Checklists are often useful during early stages of start-up operations - in large industrial systems and in preparing vehicles, including preparing aircraft for flight. Typically, the operator

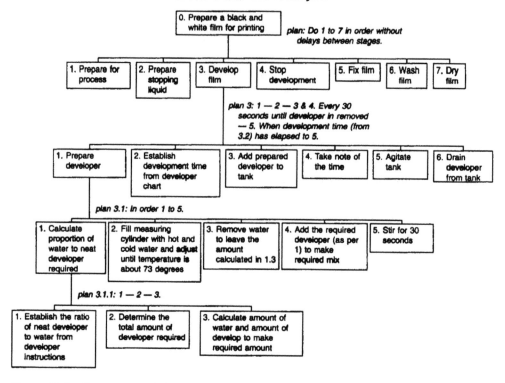

Figure 11.1   Preparing and developing a black and white film.

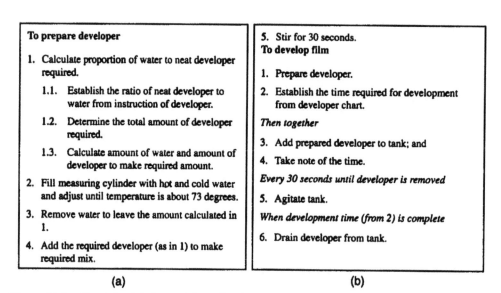

(a)   (b)

Figure 11.2   Two job-aids derived from the HTA in Figure 11.1. (a) shows a fixed procedure. (b) shows a contingent procedure.

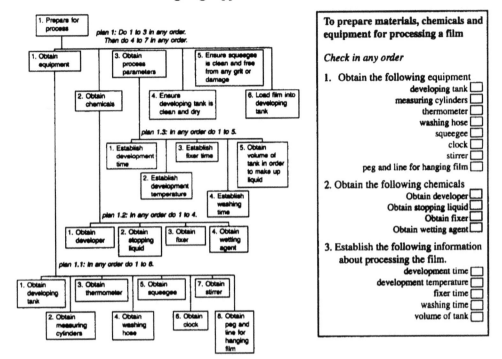

Figure 11.3   A checklist derived from HTA.

must make sure that necessary services are available and that the system is able to · function in the different ways that will be required later on. Another common feature in a checklist is that a system is left in a specified state ready for later actions to take place as planned, for example, a set of valves are left in appropriate open and closed states so that later flows of materials go where they are supposed to go. Figure 11.3 show a checklist for part of the preparation of the photographic development task. In this example, the wording from the HTA has been changed slightly to make it more accessible for the user.

Often, a record should be made to show that items in a checklist have been properly examined. Therefore, the checklist can be extended to include boxes for ticking or for recording further information, such as the time when the check was made, the identity of a piece of equipment inspected, an estimate of its state and, possibly the signature of the person making the check. These measures are certainly advised in hazardous industries. Checkboxes are also very useful for the operator to provide a record of what has been done in a lengthy sequence, especially where the checking operation is interrupted.

Decision aids

For some tasks, the problem is not so much *when* to carry out the next action, but *which* of several alternative actions should be carried out. Thus, decision aids are provided to help people make decisions and choices.

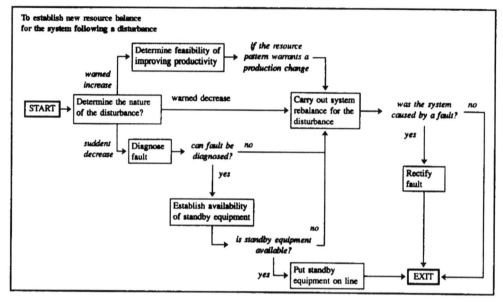

Figure 11.4   A decision flow-chart derived from the plan in Figure 3.18.

Decision trees or flow diagrams are diagrams which take the operator through a series of questions, directing which question to ask next, contingent upon answers to previous questions. Thus, decision trees chart the flow of questioning until a solution to a problem is reached. Often, it is possible to derive decision trees directly from plans in HTA, either individually or combining several plans. Figure 11.4 shows a flow diagram derived from the plan set out in Figure 3.18 concerned with rebalancing the chlorine system.

Records

When we think of keeping records we tend to think of the record keeper rather than the person who uses the information recorded. But where keeping records is justified, there is always someone who needs that information for some purpose - this might be the person making the record or a colleague. When designing records, therefore, it is important to consider why the information is required and who will use it.

Keeping records is important in many systems for purposes of accounting and to guide future action. Where product deviates from specification in an industrial process, managers and engineers will seek out recorded information in order to determine what might have gone wrong - this includes information automatically recorded and logged by instrumentation, information that operators have recorded according to a given template and information that operators have recorded in an unstructured log where they record anything they feel is significant.

The job-aid shown in Figure 11.2 (b) contains time cues, where the operator is required to take note of the current time in order to estimate when the developing tank should be agitated and when it should be emptied. In many industrial processes and care tasks, it can be critical that a subsequent action is taken at a precise time following a previous action. This could be concerned with dealing with a chemical reaction at a crucial stage or concerned with giving a drug. Industrial operations undertaken at the wrong time can damage the product or even place staff at risk. Drug treatments administered inappropriately will damage the patient. It is hard enough for an individual operator to remember when things have to be done, especially when engaged in other duties. When the time delays extend into a later shift, it is even more crucial that times when things were done, or the time when the next thing should be done, are recorded in an appropriate manner.

In some cases, records are kept to help staff engaged later on to take account of significant factors when making decisions. In neonatal intensive care, for example, when staff are planning how to give a treatment or deciding how a treatment procedure should be administered, knowledge of previous effort in administering this and other treatments can influence what a doctor does presently. For example, knowledge of actions that grossly disturbed the baby can cause the doctor to seek alternatives or take special care. When diagnosing problems in many systems, knowledge of earlier events can help reduce the operator's uncertainty about what has caused a current state of events. Devising such records and ensuring they will be used properly is not simply an issue of graphic and textual design. The designer must take heed of how activites that are carried out later are dependent upon its information. This may prompt records to be designed such that information can be recorded routinely or it may prompt the need for careful training to ensure that operators recognise when something of future significance has occurred and how best to record this information for their colleagues. It means that analysis needs to be carried out on the tasks of the people who will use the information as well as the tasks of the people making the record.

Manuals versus job-aids

Most of the forms of job-aid so far discussed are provided to support specific operations. In many cases, the job-aid is provided with the expectation that it will be used routinely. In the pharmaceutical industries, for example, there are generally requirements that detailed documentation is provided, for the operator to follow. Companies are given licenses on the condition that procedures are carefully followed to ensure that manufacturing complies with a specified standard. Many routine tasks involving computers are also designed around the principle that the user will be prompted, step by step, in the actions to be taken. For example, in supermarket checkout operations, the operator is taken step by step through the payment procedure. In hazardous industries, where certain difficult problems occur infrequently and unexpectedly, it is recognised that insufficient practice will have occurred to ensure that the operator is familiar with what needs to be done and cannot be relied upon to perform from memory. So a job-aid is provided and the operator is expected to use it. Where there is an assumption that the operator will use a job-aid, this may be used to reduce the training provided.

In many situations where operators are expected, generally, to perform unaided, they may be provided with reference documents to refer to at times of uncertainty. These documents are often called 'manuals'. Thus, people using computer applications such as spreadsheet packages may refer to the 'user manual' if they are uncertain of what to do next. An operator in a process plant may make occasional reference to an 'operating manual' to help recall a procedure which is carried out infrequently. A systems programmer will need, from time to time, to refer to a 'systems manual' when maintaining computer code, because the manual specifies how and why, things were done as they were. A maintenance technician may refer to a 'repair manual' which sets out procedures and diagnostic aids for a wide range of equipment, some of which the technician may never have encountered before. In all of these examples, the user or operator uses the manual for reference when the need arises, but is not normally expected to use the document continuously. Manuals may also be used as part of a training option to help people learn to carry out a set of tasks; if circumstances permit, the operator or user may use the manual to deal with unfamiliar operations, with the hope that something will 'stick' for a later occasion.

Manuals can become bulky. They may run to several volumes and become too large and heavy for practical use. It may also be difficult to locate the required information. HTA can be helpful in designing manuals both in ensuring that they are concise and in helping users to find their way around. Pages of manuals may, individually, look like the different sorts of job-aid that we have already discussed. The examples of job-aids that have been described have all taken advantage of the ways that operations and plans are recorded. To develop manuals, we can add the hierarchical structure of the HTA. If we consider the task of developing a black and white photographic film, set out in Figure 3.6 and extended in Figure 11.1, the top row of the analysis can be adapted to serve as a general guide to the manual, referring to the kinds of job-aids seen in Figures 11.2 and 11.3 on other pages. The hierarchy enables the designer to cross-refer all parts of the manual as shown in Figure 11.5.

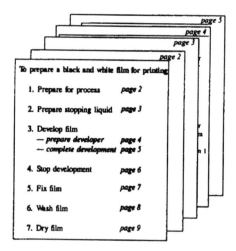

Figure 11.5   The hierarchical aspects of a task description used to aid the organiosation of a manual.

Determining where job-aids should be employed

As the task analysis is developed, the analyst must decide whether operating skills are best developed by training or whether a job-aid should be provided to support performance. Proper production of documentation can be expensive and should not be undertaken unless it is justified. There are several features that the analyst should consider in making such a justification. These include the need to conform to standards and the need to minimise operator error.

HTA does not explicitly help the analyst to make these decisions, but the analyst needs to become alert to these issues whilst engaging with the task.

The need to conform to standards

Sometimes, manuals have to be developed to comply with a formal requirement imposed by law, a professional association or company policy. In some companies it is policy that user/operator documentation is provided for all tasks. This may simply be a general house style, but often such a policy is an insurance measure designed to demonstrate to outside inspectors that the organisation is doing all that is necessary to support safe and reliable performance, for example.

Mandatory documents are sometimes necessary to comply with an industry standard. Good manufacturing practices in the pharmaceutical industries or food-processing industries, for example, often require that detailed documentation is provided to ensure that materials are produced to an acceptable standard if the product is to be used as feedstock elsewhere.

Minimising error

A major reason for developing documentation is that it will help support performance that might otherwise be prone to error. In some tasks, conditions of performance are such that the user/operator is unable to perform unaided with sufficient reliability.

Generally, job-aids may be warranted if:

- tasks may be forgotten because they are required infrequently, for example emergency procedures, specialist computing functions
- tasks are difficult, for example, diagnostic tasks
- operations to be carried out in similar but different conditions may be confusable
- procedures are long and difficult to remember
- mistakes are unacceptable
- strict adherence to procedures is necessary, for example in carrying out medial tests.

For any particular case, a combination of these factors will apply. The designer must judge the implications of any such combination.

Caution in using job-aids

A further consideration in advocating a job-aid is the time required for the operator to locate and use the aid. In some emergency situations conditions develop so quickly that rapid response is required. Use of a job-aid may be ill-advised, even though other conditions seem to justify this approach. Also, there is no guarantee that people will conscientiously use the job-aids. Even if operators are required to record 'checks' to indicate that different steps have been completed, there is no guarantee that they have conscientiously read each step in the aid. One factor which may deter use is that the job-aids may be seen by an operator as de-skilling the task.

A final problem is that job-aids are, generally, inflexible - they prescribe actions according one state of affairs that the designer assumes will prevail. A particular problem that relates mainly to decision aids is that they often only represent a small range of conditions. Thus, in a fault diagnosis task, the operator will be supported in distinguishing between those faults that the designer identified previously and will not support the operator who must deal with an unforeseen problem. If unforeseen circumstances must be dealt with, then the use of decision aids with finite fault-sets is not advised.

Relating job-aids to training

Job-aids are developed to support the performance of specific aspects of tasks. Using HTA to specify the task, it can be seen that the job-aid helps the operator deal with aspects of plans - specifying the order of doing things, the conditions to be identified to cue when things are done, or the alternative actions that are followed under different conditions. This means that there are other parts of the task that must be learned by the operator.

Chapter 6, in relation to the discussion of batch processing, described the use of 'manufacturing instructions' in helping an operator use a set of common skills. In batch processing, the same plant can be used in different ways to manufacture different products. Each product would have its own manufacturing instruction to guide the operator in using the skills with which the operator has been trained. The idea of an manufacturing instruction is like a recipe. It should refer to different ingredients, then specify the skills and conditions that need to be deployed to make the specific product. Many manufacturing instructions in industry are bulky and include more detail that is necessary for the task. They often combine descriptions of procedures that the operator practises frequently, as well as the specific batch information that is needed to distinguish the product. The consequence is that these instructions become bulky and operators may skip bits, especially those bits with which they feel familiar.

HTA can be used to develop a much more concise form of operating instruction in a way which also points to the core training that the operator needs to use a family of such instructions for a plant.

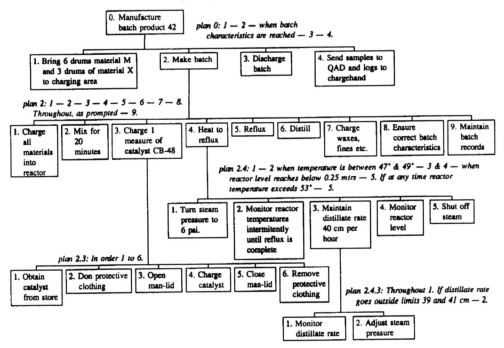

Figure 11.6   HTA of manufacturing a specific product in a batch plant.

The suggestion for analysing a batch plant operation, discussed in Chapter 6, was that the analysts should first focus on the manufacture of one particular product. Figure 11.6 shows such an HTA concerned with manufacturing a synthetic resin called Product 42. If the analyst were then to conduct an analysis of, say, Product 37, it would become apparent that much of the detailed procedure is the same. The differences lie with the type and amount of raw materials, the target temperatures and timing etc. Further comparison with this or the analysis of further products would confirm these sorts of difference and also show that for some products there may also be variants in some of the procedures themselves. An important one that emerged was that for some products it was appropriate to add both raw materials together (*en-masse*), while for other products it was essential that the products are added in a particular order, with the second being added *gradually* to avoid the risk of explosion.

Having completed one HTA, the examination of further products is done simply by charting differences. This is quite an economical method. From this exercise the list of similarities and variations quickly becomes apparent. Thus, Figure 11.7 shows a manufacturing instruction for Product 37 contained on a single page. To use this sort of job-aid, the operator needs to be skilled in each component of the task. Figure 11.8 shows the generic HTA for all products, which has been adapted to emphasise reference to the manufacturing instruction, rather than setting out specific details of batches. If the operator is trained in accordance with this HTA, then he or she can be supplied with any of the batch manufacturing instructions and be expected to perform effectively.

```
┌─────────────────────────────────────────────────────────────┐
│          Batch Manufacturing Instruction for Product 37       │
│                                                               │
│    1.  Transport 5 drums of M35 and 7 drums of N56 to charge area │
│        and catalyst C5, waxes and fines.                      │
│                                                               │
│    2.  Charge vessels                                         │
│         •  using MASS ADDITION                                │
│                                                               │
│  When lab report shows formulation correct                    │
│                                                               │
│    3.  Charge catalyst                                        │
│         •  Use respirator when man-lid open                   │
│         •  Before measuring ph, ensure temperature is in range 56 to 58 │
│                                                               │
│  When ph is correct                                           │
│                                                               │
│    4.  Heat to reflux                                         │
│         •  Turn off steam @ 67 to obtain vessel temperature of 75 │
│         •  Record reflux start time                           │
│                                                               │
│  When reflux has started at around 75                         │
│                                                               │
│    5.  Monitor and maintain reflux                            │
│         •  Maintain reflux for 3 hours                        │
│         •  Maintain temperature @ 75                          │
│                                                               │
│  After 3 hours reflux                                         │
│                                                               │
│    6.  Distill                                                │
│         •  Ensure distillation rate is no more than 40 cm per hour │
│         •  Monitor distillate receiver every 15 minutes       │
│                                                               │
│  When vacuum of 910mb has been held for 15 mins               │
│                                                               │
│    7.  Charge waxes and fines                                 │
│         •  When charging is complete, bring temperature back to 95 │
│                                                               │
│  When temperature is 95                                       │
│                                                               │
│    8.  Ensure viscosity is within limits                      │
│         •  Use lab procedure 27.                              │
│                                                               │
│  When viscosity is within limits                              │
│                                                               │
│    9.  Discharge batch                                        │
│                                                               │
│  When batch is discharged                                     │
│                                                               │
│   10.  Send samples to QAD & logs to chargehand               │
└─────────────────────────────────────────────────────────────┘
```

Figure 11.7   A batch manufacturing instruction for a single product.

It is important to stress that once a job-aid has been developed, the operator must be trained in its use. The operator must become familiar with the job-aid, learn how to use this in the context of the job and gain confidence in using it. This example shows the benefits of considering training and job-aids together in solving practical performance problems.

Representation of job-aids and the link to HTA

To conclude, we can revue some of the general guidance that is often presented to help in the development of documentation and to show how using HTA complies with these recommendations. It is not appropriate to develop these here.

Presentation is an important issue in developing job-aids. It deals with layout, choice of font, text style and text size. There are some important principles and guidance to be followed in designing job-aids, whether these appear on paper or on a computer screen. What must be emphasised, though is that issues of presentation are as much

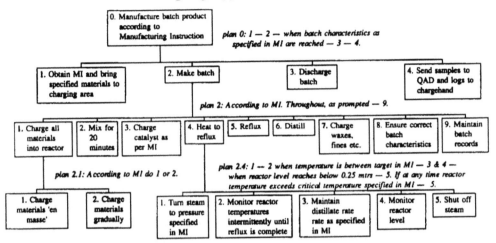

Figure 11.8   A generic HTA for dealing with all products on the batch plant. Plans have been revised to refer to the Manufacturing Instruction for information

concerned with issues of consistency as aesthetics. In the various job-aids presented, different text styles and layout have been used, but care has been taken in ensuring that different styles refer to different sorts of thing. For example, the use of italics has generally been reserved to indicate cues that the operator or user should attend to, whilst plain text has been used to represent actions to follow.

Ensure that the words used can be understood. Use words which match the vocabulary of the user/operator. If colloquialisms are used consistently in the workplace, these should be adopted in the written documentation (good taste permitting of course!). If there is an inconsistency in the way in which terms are used, by different personnel who must interact, for example, then these inconsistencies must be resolved by appropriate training. Technical terms or those with a possibly ambiguous meaning, should be included in a glossary of terms to which the user/operator can refer. In developing and HTA the analyst is always advised to describe operations using the terms that are used and understood by operating staff.

Instructions should be expressed in the active rather than the passive voice. An instruction expressed in the *active voice* might say, 'start the pump', whereas that in the *passive* voice, would say 'the pump is to be started'. It is far easier for the reader to understand an instruction in the active voice and convert it into action. The manner in which operations and goals are recorded in HTA - in the form of simple instructions - means that operations in the HTA can be translated easily into instructions within a job-aid.

Use negatives with care. Sentences which avoid the use of negative words (such as 'no', 'not', 'never') are easier to interpret than sentences which include them. The simple assertive phrasing favoured in HTA makes translation straightforward.

Use short concise sentences with minimal information content. Sentences used in instructions should not be long-winded. The user needs to read and commit the instruction to memory, then follow the instruction. If the instruction is too long, steps will be forgotten and errors made. HTA normally represents the task into simple steps that can translate into simple instructions. The plan governs how these steps are linked.

Avoid ambiguities - be explicit. There are two issues here. First, HTA provides a way of describing each operation in a task context, such that the instructions derived for the job-aid are most likely to conform to the logic of a real task. The second issue is not to omit key words such as 'the', 'a', 'that', 'which', 'of', 'for' etc. when providing a job-aid. It is often tempting to reduce the length of text by such words with the risk that it creates confusing strings of words which cannot be understood. For example, 'Undo safety cover protecting switch' could mean 'Undo the safety cover which protects the switch', or it could mean Undo the safety cover while protecting the switch. For brevity, such short-cuts may have been employed in the HTA. It is important that the design ensures that this shorthand is not carried over into the job-aid. The HTA is *not* the job-aid; it is merely the basis from which the job-aids is derived.

Write down instructions in the order in which they must be carried out. If more than one instruction is given in a procedure, they should be mentioned in the order in which they are to be carried out. In the photography task, the operator should be told to prepare materials and equipment before starting the processing and not told, for example, that the washing facilities should have been prepared at the stage when they are about to be used. HTA makes absolutely clear what should be done and when. Such problems should not arise if the HTA is done conscientiously before job-aids are developed.

A related issue concerns the relationship between cues for action and the actions themselves. The cue for any action should be written first to alert the operator what to look for. If the cue follows the action, then the user might be focusing on the action and may have even started doing it before realising that it was necessary to wait. Recording cues before actions is a useful discipline within HTA. If it is followed, then it will also translate to writing job-aids.

Concluding remarks

This chapter has set out the considerable parallels that exist between HTA and writing job-aids. HTA represents tasks using explicit statements that link action to conditions. Thus, HTA can be used as a basis for developing job-aids. By first carrying out HTA, the job-aid designer can be reassured about the nature of the task and how it is carried out. The task can then be inspected to ensure that other design issues are optimised. Then the analyst must judge the extent to which performance of different parts of the task will be facilitated by the use of a job-aid. When these decisions have been made the analyst can identify training content and job-aid content. Job-aids are then developed and used in the context of training to help people carry out their duties. HTA contains many of the features that any job-aid designer is encouraged to incorporate in a job-aid.

Chapter 12

Human resource management issues

Human resource management must be effective if systems are to be properly staffed. These include identifying and specifying the jobs to be staffed, identifying manning requirements to ensure that sufficient personnel are recruited identifying selection requirements and specifying selection methods, staff training and staff appraisal and development. HTA can complement these by providing an analysis of the task which has already been derived for the purpose of supporting human factors decisions. Doing this will help ensure the consistency of decisions between human factors and human resource management.

Introduction[51]

Once a system has been designed and built, staff must be recruited, prepared for their duties, then supported and developed. These responsibilities fall within the remits of different professional groups who may be variously referred to as human resource management, personnel management or organisational development management. These professional disciplines employ a substantial armoury of research, principles and methods. The term human resource management will be used to refer to all of these disciplines as they all serve a similar function in maintaining the human resource, even though they may focus attention on different aspects.

It is important that human resource management decisions are based a proper understanding of the jobs they are required to support. Furthermore, taking account of task analysis already completed to serve the requirements of *human factors* design is both economical in terms of effort and promotes greater consistency between solutions.

There is also a need for managers and other professionals with different specialisms within organisations to communicate with one another. Thus operations management needs a means of representing their requirements to human resource management in order that the latter can fulfil their duties. Equally, human resource management requires methods for eliciting requirements from operations management. HTA can support these communications requirements in a useful way which is both valid and economical. Hence, this chapter will first discuss the general issue of basing human resource and personnel decisions on task analysis and then it will review how HTA lends itself to supporting a number of decisions that human resource personnel must take.

Establishing a proper task description

An important observation from applying task analysis methods, is that they reveal aspects of jobs that were not appreciated from the outset and which are often not appreciated by the staff who are familiar with them. The examples of this are numerous. For human resource management it is essential that a proper analysis of needs is undertaken. Otherwise, the wrong people will be recruited and training and development will be inappropriate.

There are many cases where the *apparent* attributes of staff are not borne out by what they actually do. Hospital doctors see a clear demarcation between what they do and what nurses do. Their view is often that doctors diagnose and nurses nurse, which means that nurses mainly carry out care routines within the parameters set by doctors. Yet any task analysis undertaken of the activities of nurses shows that they also need to diagnose, both to judge what they should do themselves and judge the extent to which they refer problems to doctors. If this were not the case, doctors would soon complain because they would have too many problems referred to them.

Process control tasks in the chemical industry are often assumed to be concerned mainly with understanding chemistry and physics. This is reflected when one sees training programmes dominated by explanations of simple chemistry and physics relating to the process. Yet, in most analyses of these sorts of tasks, such chemistry and physics is insignificant in comparison to other issues. Process controllers need to know something of *chemical engineering*, to know how materials are moved around and the risks associated with doing this. But they only need to know as much as the task requires - they do not have to be chemical engineers. Knowing something about chemical reactions can be helpful in understanding the significance of certain safety procedures, such as the use of antidotes to poisoning or various limitations on how different sorts of fires can be treated, but such knowledge is rarely directly relevant to making decisions about controlling the process itself. Generally, though, they must learn procedures and learn to make decisions - and the extent to which they do these things will vary from company to company and department to department.

In many jobs, dealing with people is frequently overlooked in favour of overemphasising the technical content of jobs. Field service engineers, looking after computer and domestic appliances for example, often need far less detailed engineering knowledge than people who held similar jobs a few years ago. This is because such equipment is now often modularised. Faulty modules can be easily detected and replaced at site, then thrown away or repaired in a workshop by specialists. The field service engineer's job focuses on moving equipment around, following well thought out procedures, interpreting instrument readings, replacing items and dealing with members of the public. The skills in actually making a repair (though not deciding which repair to make) are often similar to replacing a printer cartridge, as in Figure 6.1, although there will be some general skills concerned with pulling out old parts which have become degraded. There are skills concerned with movement and storage of equipment. In some computer installations there is a need to appraise whether floors can take the weight, whether a door will give access and whether additional thermal control is necessary - all of these depart from the initial expectation of what the engineer might be required to do if attention is paid to the product of the task. Dealing with members of the public is crucial. Engineers called out to make a repair are often representing their organisations in circumstances where customers are discontent, so their skill in handling customers is paramount.

Task analysis of any small business is revealing. Market traders may be seen as people who stand in the open and shout about the things they have to sell. People running small confectionery shops, tend to stand behind their counters in a far more sedate manner, responding to requests from customers who come into the shop. Craftsmen concerned with making antique musical instruments for specialist orchestras or museums are thought of in relation to applying fine skills to raw materials using unusual tools. Yet carrying out HTA of each of these jobs produces surprisingly similar results, at the general level at least. Each of these people runs a small business. Each must obtain and manage premises, attract customers, identify what their customers require, procure materials to sell or from which to manufacture products to sell, store raw materials and product in suitable conditions, transport materials, transact sales, maintain accounts, and deal with public agencies, such as the Inland Revenue, Customs and Excise and local authorities. Indeed, such businesses need to support the same general functions as larger businesses employing several people. In a small business, though, many or all of these functions are carried out by the same person.

There are many important aspects of jobs that will not be revealed directly by an HTA, although using HTA can draw attention to these factors anyway. For example, the requirement for shiftwork may be identified as handover tasks are identified. The stress levels in the tasks may be revealed when confronting the range of things that people do. The extent to which people work on their own or engage with others will become apparent as the task is observed and analysed. The extent to which work is dirty should also become apparent. So, while these aspects may not formally emerge from the HTA, they are likely to become apparent to the analyst.

Making human resource management decisions

When we turn to making human resource decisions themselves, details emerging from the HTA are often directly useful, for example:

- identifying and specifying jobs
- identify manning requirements
- identifying personnel selection criteria and devising selection methods
- preparing job descriptions
- training staff
- appraising and developing staff.

Identifying and specifying jobs

Providing staff and giving them adequate support requires that the nature of jobs is properly appreciated. This means understanding the skills required of personnel, the demands with which they are expected to cope and the conditions in which they are required to work. These are all issues to which task analysis methods can contribute. There are also important social, political and strategic dimensions to many job-design decisions, for example to create collective ownership of work and enhance joint decision-making and responsibility for work done. Sometimes jobs are created to provide opportunities for certain people to gain status or experience. This may be done to ensure a suitable route for people to progress through the organisation such that future skill-levels within the organisation can be assured, i.e. manpower planning. Task analysis can contribute to these social, political and strategic decisions by providing the building blocks from which actual jobs can be assembled. Thus, by using HTA the human resource professional has a basic statement of the task which will fulfil operational needs, which can then be used to satisfy the social and political dimensions as well. Issues of job-design were discussed in Chapter 8.

Identifying manning requirements

Companies often adopt manning levels according to long-standing experience of similar operation, manpower planning implications and financial limitations. They must also take note of shift requirements. The human factors perspective should also be taken into account since it emphasises the need to appreciate workload, the complementary nature of skills within particular jobs, the compatibility of skills between jobs with a working team or group. These issues can be investigated by reference to task analysis methods. Ultimately, manning requirements must be evaluated against the requirements of the task and the need to respond to typical and critical events. These issues were also discussed in Chapter 8.

Identifying personnel selection criteria and devising selection methods

Recruiting for jobs requires drawing up a specification of what staff are expected to do. HTA tables have been used to review design options within tasks. They can be used in this way to consider selection criteria. In particular, it is useful for people managing the recruitment process to review the task analysis to identify which aspects of skills can be trained and which aspects would benefit from recruiting people who already possess the required skills and attributes for the task. Table 12.1 shows the task of a health physicist. Health physicists are employed in nuclear installations to monitor the health of employees and levels of radiation around the site and then to investigate problems identified. The job entails technical knowledge concerning the nature and processing of nuclear materials and the affects of radiation on health. It also requires the job-holder to work with an approved set of parameters to ensure adherence to proper procedures. Analysis of the task reveals several operations in which this technical knowledge is used to investigate problems. It also shows the extent to which the job-holder needs authority, confidence in judgements made and considerable organisational and supervisory capabilities. Table 12.1 has been used to scrutinise the various parts of the task to consider which attributes a job applicant would need to possess and how these attributes can be investigated.

The table also shows different ways of exploring the attributes required. Evidence about applicants will be obtained by reviewing CVs and references, operational interviews to establish technical competence, case study/simulation exercises to assess decision-making and judgement. Selection exercises can be derived from the HTA, for example, in-tray exercises can be devised by identifying the range of information that the job-holder will be required to handle and the decisions to be taken. The information and decisions that are appropriate for including within the in-tray exercise are then selected and items for the exercise are then devised. Other selection exercises can be devised to simulate the essence of the tasks identified through the HTA.

An important aspect of any selection method is that it is validated, that is, it can be shown that applying the method will help personnel make appropriate recruitment decisions. In essence, validation entails demonstrating that people selected will go on to perform well in the job[54]. This requires that, once recruited, assessments are made of how people develop in order that the recruitment methods can be evaluated and revised as necessary. Where substantial numbers of people are recruited for a position and, especially where psychometric methods are employed, validation is assessed formally by correlating selection scores with scores of job performance. In many situations, however, such evaluation is impossible. This is particularly the case in complex and varied jobs, such as those where people must respond to different circumstances, rather than carry out straightforward procedures. In these cases especially, HTA can be used to advantage to assess the actual tasks and skills involved in jobs. On the one hand, this

unfamiliar with the new technology and have no experience of the new working arrangements of the team. These skills, therefore, are not equivalent to what supervisors in the organisation already do, so the recruitment pool may need to be widened. The third duty concerns liaising with management to account for the team's performance and dealing with incidents. The fourth duty requires that the supervisor is able to undertake all of the duties of other team members - again, these will be new demands on anyone already in the organisation. The job description is derived from the HTA; the HTA is still available to explore the nature of these duties further.

Training staff

Training is a discipline that sits between human resource considerations and human factors considerations. Human factors is concerned with training to the extent that it relates to operator skills as they need to be deployed to meet a system's goals. The human resource perspective is concerned with issues such as staff development and qualification. Of course, the two perspectives relate closely to each other. Qualifications of staff need to be relevant to performance needs and so, identifying appropriate qualifications for staff should be based on the analysis of the tasks they have to carry out. If this is not done, people acquire qualifications that are not strictly pertinent to their needs. Chapter 10 set out the relationship between HTA and training with key aspects summarised in Table 10.3. Here, the assessment of training needs is particularly relevant to personnel decisions.

Appraising and developing staff

Once trained all staff should be monitored to ensure that their skills are maintained and also given opportunities for advancement. Staff appraisal can be assisted considerably by a proper understanding of what staff are employed to do. The HTA will provide a proper justification for listing the criteria against which judgements are made. Staff development is concerned with giving people new opportunities in which to explore new skills and responsibilities. This entails identifying tasks which can be delegated to staff being developed, then supporting and appraising their efforts. This amounts to the process of delegating to a trainee, set out in Figure 8.9.

Concluding comment

Human resource management is a substantial discipline dealing with many issues. If an HTA is produced in a system in order to support human factors design, then it can also be used to support a number of personnel decisions, concerned with recruitment and support for personnel. The benefits to be gained from doing this include providing a focused description around which human resources staff can develop their ideas in a way which reflects the requirements of the system. It also provides a common task description from which a number of different human resource decisions can be derived. This is economical and it will also ensure consistent results which are compatible with the decisions being made in human factors design.

Chapter 13

Conclusions

The book has described the processes of HTA and shown a variety of ways in which it can be applied. In both respects, the discussion has been developed from a practical perspective to support the view that task analysis methods should be regarded as practical methods to help with the processes of human factors and human resource management.

The method

HTA was developed as a practical tool to facilitate human factors decision-making. In justifying their approach, its authors used concepts from skills psychology - goals, planning feedback - along with cost-benefit analysis in order to help to make the method sensitive to the context in which it was being applied. Tasks were analysed by examining the operator's *goals* and the *operations* they carried out on the system they were dealing with in order to meet these goals. Thus, goals related to the operator's intentions to the needs of the system in which they were employed. Strategy and behaviour may change from occasion to occasion, and the competent operator will adapt to circumstances to ensure that the goal is met. Yet the operator is still assumed to be carrying out the same *operation*.

There will be occasions when the analyst judges that it is necessary to gain further insight into how the operation is carried out. In HTA this can be accomplished in two ways. One of these is by examining the *behaviour* involved in carrying out the operation. A simple method for accomplishing this is the I-A-F model, in which performance problems are accounted for in terms of whether they affect the operators capability to deal with the cues to prompt action (*input*), the capability to carry out the *action*, and the interpretation of the *feedback* to regulate the action. While this was originally advocated as part of the HTA process, there is no reason why other methods should not be used, including *cognitive task analysis* methods. Indeed, this is an effective way of linking HTA to other methods.

In HTA, as in all task analysis methods, speculating about the behaviour involved in an operation can only ever lead to hypotheses about performance and design and cannot lead to certainties. This is because all tasks may combine novel elements whose interaction in the task context cannot be anticipated with complete confidence. All ergonomics and human factors methods which attempt to assert how people behave in different circumstances share this feature, although for some issues it is possible to be more confident in a hypothesis that for others. Hypotheses need testing, either by conducting formal trials or observing the outcome of decisions made in a task analysis as they are applied in the real world. This is a particular problem when making inferences about psychological processes in complex applied situations where a range of different strategic behaviours is possible. The hypothetical nature of making design decisions is acknowledged in most design methodologies, where evaluation and revision of design specification is regarded as inevitable. In such systems, the aim of making a design specification is to make choices which *minimise* redesign as much as possible.

The second method, used in HTA for gaining further insight about a task, is *redescription*. Here, the analyst seeks a set of *subordinate* goals and a *plan* which governs when these subordinates are carried out. It is the repetition of this process that develops the hierarchy of goals and plans that characterises HTA. Plans have an important function in accounting for the complexity of tasks. The role of plans has featured centrally in this book.

A further important component of HTA is *stopping redescription*. If the analyst can be clear idea about how much detail is necessary, then unnecessarily large task hierarchies can be avoided. The main stopping rule published in connection with HTA is the P x C rule, where P represents an estimate of the *probability* of inadequate performance and C represents the *cost* of inadequate performance. Thus, the decision to stop redescription is taken with reference to *risk*. The P x C rule, in this simplest form, is misleading because it implies that the analyst will formally assess these values, which is rarely possible in practice. This is particularly the case when estimating, for example, the cost of an injury or even a life. It is also the case that other factors come into play with regard to this *cost-benefit* analysis, for example the costs of alternative design solutions - there is no benefit in avoiding an operational cost if it is exceeded by the cost of the innovation. Also, organisations who limit their cost-benefit analysis to the immediate task are unlikely to be innovative. Investment in information and communications technology, for example, is usually based on a longer-term perspective than consideration of a particular system in a particular context. Finally, factors concerned with cost-benefit analysis are not the only considerations in deciding whether to stop or continue with redescription. In designing displays for example, it is often most helpful to continue redescription until operations are expressed in terms of interaction with the instruments and the controls that operators will use. Perhaps the most important guideline, with regard to stopping further redescription, is that the analyst is clear about why the task analysis is being conducted. This wider purpose will influence the judgements that the analyst makes along the way.

Initially, HTA was justified by reference to a hierarchical model of skill, where plans were treated as methods to select subordinate control routines. These sorts of model are problematical as a basis for task analysis because they appear to suggest that the hierarchical description obtained represents how operators really organise their thinking and planning. The analyst cannot make any real justification of this. The methods used to examine performance cannot categorically state that a particular hierarchical method of organising the task is used by all people called upon to perform the task, or even used consistently by an individual carrying out the task on different occasions, or even that performance is arranged hierarchically at all. Indeed, human behaviour is far more adaptive than that implied by assuming that HTA is a straightforward representation of how different people organise their behaviour. People modify their strategies with experience, in accordance with how urgently or critically they must deal with tasks and in terms of the effort they must make to obtain task information. Thus, in Chapter 2, justification of HTA in terms of modelling behaviour was avoided. Instead, HTA was described as a process in which a task analyst engaged with a problem by representing it in terms of its main goals then considering how it related to its wider context. The analyst then investigates the problem by examining the operator-task interaction and then by redescribing goals in greater detail using subgoals and plans. The outcome of this process is the hierarchical description that characterises HTA.

The power of HTA rests substantially with the power of plans. Chapter 3 showed the range of plans that can be encountered and that relatively simple plans can be combined to account for apparently complex activities. In this way, the analyst is able to be thorough in focusing on different parts of the task.

A simple representation of HTA, given in many publications, is that it is a process of redescribing goals in terms of subordinate goals and plans in order to describe the task as a hierarchy of goals. In one sense this is reasonably accurate, but it fails to represent the assumptions and process involved. Such a simple description is easy to misrepresent in terms of assumptions about human behaviour that cannot be justified - in a practical project, at least.

The application of HTA to different domains

Many published accounts of HTA and other task analysis methods are limited to short descriptions in textbooks. Often the examples presented are simple tasks to illustrate the concepts rather than to demonstrate the scope of their application and are often represented to very few domains. Chapter 6 provided examples of HTA from a wide range of domains including, domestic tasks, medical tasks, computing tasks, tasks concerned with production and maintenance, and tasks concerned with management and supervision. HTA in every one of these domains shares features encountered elsewhere. An analyst examining an unfamiliar task in a new domain is able to try out task structures observed elsewhere. Thus, monitoring a child's health in intensive care has similarities with monitoring a process plant or an air-traffic sector. By applying these task representations from other domains, the analyst can often make progress. Then, by applying the rules of HTA, the analyst is able to see where the new task is different. Indeed, caring for a baby in an intensive care ward is substantially different from running a nuclear power plant even though there are also some similarities. Similar structures can be repeated throughout a particular task analysis to account for its complexity and these structures can be seen to apply to HTAs in different domains. Chapter 3 dealt with the variety of types of plan that can be encountered in different analyses.

Efficiency is gained as the analyst becomes more expert. A more experienced analyst can more readily recognise how elements of a task can be represented, by trying out task structures encountered elsewhere. The basis for this comes when the analyst becomes more versed in understanding tasks in terms of their functionality and is able to make reference to more and more different tasks in different domains, to see how similarities might be applied. One of the aims of the book has been to provide many examples of HTA from different domains, so that the reader is better able to make links.

The application of HTA to supporting different design solutions

Writing on task analysis has often failed to show how it can be applied. HTA and many other task analysis methods are tools to serve in the process of human factors design and investigation. Their benefits, therefore, must be judged in terms of how easily they enable insight into the task to be gained and allow the processes of design for human factors and human resource management to be carried out. Therefore, considerable space was devoted in the book to illustrate some of the ways that the HTA can be used to facilitate the design process. HTA, it is argued, can be applied to support a range of human factors decisions: team and job-design (Chapter 8); information requirements specification and work design (Chapter 9); training (Chapter 10); development of support documentation (Chapter 11); and some human resource management issues (Chapter 12). The motivation for developing these applications was in part to demonstrate how HTA can be used in practice and also to demonstrate how HTA can be used with economy and consistency throughout real projects.

In practical task analysis projects carried out in organisations, the analyst can be confronted with the view that task analysis is an intrusive luxury. This can be refuted. HTA enables the analyst to obtain necessary information and to do so as quickly as is feasible. The stopping rules in HTA are designed to ensure that analysis only proceeds where it is justified. Therefore, if the task analysis takes a long time it is because the task is large or complex. Organisations need to develop places for people to work and methods to train them, so they must engage in some effort to understand what is involved. HTA is an efficient and effective way of doing this. Further economies can be gained by making sure that the effort directed towards one design goal can be used for others. Therefore, managed sensibly, HTA can be developed systematically across the life of a system to apply to the different human factors and human resource management decisions encountered - as argued in Chapter 7. Added to these benefits of *economy* are benefits of *consistency*. It is common to encounter operating manuals and training manuals with different authors who have adopted different views about the task which are also at odds with what is actually done.

Cognition and flexibility

Many tasks entail cognitive elements of monitoring and decision-making. Many authors have argued that different methods are required to analyse cognition in tasks. However, distinguishing between forms of cognitive task analysis and non-cognitive methods makes little sense when all tasks entail cognition to some extent. The framework for task analysis, set out in Chapter 2 argues that it is best to pursue an evenhanded approach to analysis, then use specialist cognitive approaches, if necessary, as cognitive problems are encountered. This would enable cognition to be related sensibly to other parts of the task such that its precise function with regard to the execution of the task is better understood. It also places cognition in the wider task context such that the criticality of cognition can be assessed. Following this approach, Chapter 4 further argued that, in many cases, detailed cognitive analysis is unnecessary - the basic HTA method can be applied to understand where people make decisions, the nature of these decisions and the context in which these decisions are made. Often, this degree of understanding is sufficient for practical purposes.

It is ironic that one of the original justifications for HTA was to provide a method of analysis able to deal with the emerging work practices characteristic of more automated industries. Despite this, HTA is often characterised as being inappropriate for analysis of 'cognitive' tasks. HTA is not a cognitive task analysis method, but it is a method in which cognitive elements of tasks are identified and can be appraised and which provides a clearer context for using cognitive task analysis methods should this be judged appropriate.

Tasks and contexts

Task analysis presupposes a task to be analysed, yet it is difficult to pin down what different people really mean by the word 'task'. Chapter 1 provided a definition of 'task' which was, at least, consistent with how the word is used with respect to HTA. It treated a task as '*a problem to be solved or a challenge to be met ... a set of things including a system's goal to be met, a set of resources to be used, and a set of constraints to be observed in using resource*'. Goals relate to the operators' intentions and resources relate to the tools at their disposal to meet goals. The role of *constraints* is less obvious. Constraints place conditions on how resources can be used. An operator trying to solve a problem must observe constraints in deciding which resources to use and how to deploy them. Equally, an analyst working with a task expert will observe constraints in deciding how best to describe a plan. Thus, in HTA, constraints are taken into account, but there is no format for recording them independently and automatically. Chapter 5, concerning representation of HTA, suggested that constraints, including details about performance criteria should be recorded in the notes column of the HTA table. This would provide a record to show what was taken into account by the analyst. This can be important because lower levels in a hierarchy may share the same criteria and constraints with each other, as do different parts of the hierarchy which may be linked through their plans. However, this solution is not particularly rigorous and methods that better account for how constraints are recorded within HTA would be welcome. One solution would be to develop computer tools to record issues of criteria and constraint more systematically recorded so that their consequences can be referred automatically to the other parts of the task. Then they would not be overlooked when design decisions are taken or resource constraints overcome.

HTA and other task analysis methods

An aim of this book has been to make the case for HTA, to justify its method, show how task analysis can be carried out and how it can be applied to deal with design issues. Its purpose has not been to make any strong comparison between HTA and other task analysis methods. It is appropriate to comment on this now.

It should be apparent that HTA cannot be conducted on its own without reference to other task analysis methods. The analyst will engage, from time to time, in formal and informal interviews, observations, scrutiny of performance records, simulation exercises, verbal protocol analysis, time-line analysis, link analysis, withheld information exercises, cognitive task analysis and so forth. The analysis will also employ a range of representational methods, including, flowcharts, rule-sets and tables. Figure 2.3 set out the cycle of task analytical decisions that lead to the development of the HTA. To make these various decisions entails obtaining and organising data and this can be done using any of the specific task analysis methods that are judged appropriate. Appropriateness should be judged in terms of the opportunities afforded by the context to obtain data, by the suitability of the method and the analyst's experience and preferences. In this sense HTA may be adopted to deal with all situations, 'cognitive' or otherwise. The analyst, it is suggested, should not make decisions from the outset about which methods of analysis will be adopted to investigate specific features, but use the HTA framework to guide what information should be collected. This will then prompt which task analysis or data collection methods to fulfil this aim. The outcome can be accommodated within the wider HTA. This approach can apply even where a task is heavily cognitive or concerned with environmental issues or workspace issues. All this means is that the analytical process begins by properly appreciating the goal to which the behaviour of principal interest is directed, then progresses systematically in order to situate the specialist analysis within the context of the wider task.

As a final comment, I would like to acknowledge a general point concerning different task analysis methods. The purpose of task analysis methods is to engage the analyst with the task in a systematic way - no task analysis method or framework is helpful if the analyst merely sits back and hopes that all the mysteries of the task will be revealed. Proper insight and understanding will only arise when the analyst becomes engaged with the task. Many different methods of task analysis can assist with this engagement and many will be capable of prompting the critical insight needed to make sense of a task. It should be perfectly understandable to the reader, though, if I were to conclude with an assertion that HTA, as a framework for examining different facets of a task, provides more opportunities than most methods for engaging with the task in different and useful ways which are of relevance to any practical human factors project.

Chapter 14

Notes and references

This Chapter contains notes that have been indicated within Chapter 1 to 13 and should normally only be read in conjunction with these reference points. The notes contain references to other books and papers and the full reference list is provided at the end of the Chapter.

Chapter 1: Task analysis, concepts and terminology

1   'Hierarchical Task Analysis' (HTA) was developed by Annett and Duncan (Annett and Duncan, 1967; Annett et al., 1971; Duncan, 1974). Their work was aimed at setting out general principles appropriate to guiding all task analysis projects.

2   Systems thinking underpins much of the work in human factors. It provides frameworks for examining complex problems, methods for describing the complexity of human performance and frameworks for integrating human factors into the wider processes of design. The basic ideas of systems are summarised by Jenkins (1969).

3   That human performance can only be properly appreciated by reference to its context is fundamental to applied disciplines such as ergonomics, human factors and human computer interaction design.

4   The term used to refer to the person carrying out the task varies according to taste and fashion. It is important to adopt a neutral term because job titles can be misleading. 'Operator' used to be a standard ergonomics term, but 'user' was adopted with the development of human computer interaction research. Generally I shall use the term 'operator' to denote the person carrying out the task to reflect that view that people *operate* upon information and materials in order to achieve goals.

However, where the focus of interest is how people use a tool, such as a wordprocessor, I shall use the more appropriate term 'user' or even a specific job title. In using either term, I am merely labelling the person engaged and am not implying any degree of expertise, authority or discretion. In each case, it is the task analysis that will indicate these things.

5 A slow response system is one where the consequences of the operator's action take time to emerge, for example, raising the temperature of a liquid in a vessel takes time for everything to heat through. This contrasts with fast response systems where the consequences of action are known almost immediately. In fast response systems, such as driving, the operator may make continual adjustments in order to keep to a target. Thus, errors can be quickly detected and rectified. In slow response systems, rectifying an error is subject to lag - often, problems are manifested some time after their initiating event has occurred and then there is a time delay before remedial action can take effect. Therefore, the operator must think ahead and anticipate the consequences of any action planned before implementing the action. To select suitable responses for dealing with system disturbances, the operator must understand something about system dynamics or learn to evaluate current circumstances with a view to selecting from a set of proven suitable actions. Slow response systems are also often complex and hazardous, so the difficulties of the task due the system lags becomes particularly critical. Such system characteristics are common in process control operations (e.g. Edwards and Lees, 1974; Moray, 1997). The lag in systems causes problems in many domains, including transportation, healthcare and organisations. Operations, such as these are of central concern in task analysis.

6 Annett, Duncan, Stammers and Gray (1971).

7 It would seem likely that, in view of the extensive writing on task analysis, the word 'task' would be clearly defined. This is not the case. Generally, investigators do what they think will be useful in revealing the aspects of performance that cause problems or to develop design; they call the activity of investigation 'task analysis'. Most task analysis methods are conducted without first clearly identifying the object of analysis. It means that task analysis is generally defined in terms of its *process* and *outcome* and does not depend on a clear definition for the thing being analysed. Stammers et al. (1991) defend this view, arguing that the looseness of definition avoids placing unnecessary constraints on the analyst.

Definitions of 'task' are rarely entirely helpful. One of the most well-known is that by Miller (1953) - "a task is a group of discriminations, decisions and effector activities related to each other by temporal proximity, immediate purpose and a common man-machine output". This seems to rule out things that occur at diverse times so, by this definition, it might be inappropriate to treat management or supervision as tasks because they entail attention to a variety of general, as well as specific, goals which may occur in different contexts and which extend over long periods. The definition is also limited to focused actions. Diaper (1989a) supports Miller in limiting a task to a short duration and suggests the word 'project' is used to describe activities of longer duration. This does not really solve any problems, especially in system supervisory activities, where the operator must link different activities across considerable time periods. When a system fails, its diagnosis and plans for its recovery may use information gleaned hours days or weeks before. Imposing time limits before the task is examined may be quite misleading. Meister (1985) discusses the dimensions of tasks, whilst tacitly accepting Miller's definition. He points out a number of system dimensions - for example, temporal relations, functioning, dependence, complexity, organisation - that may need to be taken into account when addressing the issues that influence performance. The definition I have used, which refers to goals, resources and constraints, emphasises that tasks are related to system functions. This enables a range of behaviours to be adopted in meeting these goals, provided these behaviours use available resources and observe system constraints.

Chapter 2: HTA - a task analysis framework

8   The P x C rule is discussed in Annett and Duncan (1967), Annett et al. (1971) and Duncan (1974). Extensions to the P x C rule are discussed in Shepherd (1985 and 1989).

9   The I-A-F model was an integral part of the method of analysis proposed by Annett and Duncan, even though it is the hierarchical structure of HTA that has been most prominent. This reinforces the idea of HTA as a framework for the systematic analysis of tasks, rather than simply a method of representing operations and plans in a hierarchy.

10   Examination of operations provides the opportunity for the analyst to consider a wide range of perspectives. Indeed, many task analysis methods are actually formal or informal models of behaviour. Some of these can be regarded as information processing models of behaviour where the analyst considers how information from the task interface is processed by the operator in order to determine an appropriate response. R.B. Miller's (1966) task analysis method was based around the idea of an information processing model in which information is sensed, perceived, organised, then used to select a response through the application of elements of short- and long-term memory. Miller's approach was devised explicitly to support task analysis. The analyst would use the model to identify potential sources of error and then prescribe methods for overcoming the potential problems. Thus, if it was felt that a problem was likely to arise due to an overload in short-term memory,

then the analyst might suggest a modification to job-aids or to displays to reduce this memory lead. There are many other such models, not always developed with task analysis in mind but still potentially applicable. Thus, Wickens' (1992) information processing model is similar to Miller's.

Other approaches to examining operational behaviour concern the content and organisation of task knowledge. Gagne (1970, 1988) has demonstrated the merits of considering behaviour in terms of a hierarchy of cognitive elements, ranging from the identification of simple signals through to the application of concepts rules and principles. His suggestion is that mastery of more complex task elements relies on the development of less complex task elements. Rasmussen, in numerous articles (e.g. 1980, 1986) has presented a model of information processing which shows how the operator may adapt a response in accordance with issues such as problem familiarity, the representation of information, the type of training received and risk. Thus, in some circumstances, the operator will respond by simply matching known responses to presented patterns of events, sometimes, general rules will be applied where responses are made according to classes of events, and sometimes the operator will develop a response from the application of basic principles. These are, respectively, commonly referred to as *skill-*, *rule-* and *knowledge*-based methods of responding and the operator may switch between these approaches according to experience and circumstances. For some operations, operators must make decisions and plan. Thus, approaches such as TAKD (Diaper, 1989b) seek to identify task knowledge assumed to be used in making these decisions. Shepherd (1995) gives a more detailed account of several of these different approaches.

11 Wilson and Corlett (1995) contains examples of, and references to, numerous ergonomics checklists dealing with topics such as *seating design, health and safety, manual handling, stress* and the *visual environment*. An analyst might choose to examine an operation from the perspective of any one of these topics. Chapter 6 in Kirwan and Ainsworth (1992) describes the development of ergonomics checklists and provides references to a number of checklists concerned with different aspects of human factors, including cathode ray tube display/visual display unit design and use, maintenance, equipment design, human error, transport and user interface software. Clegg et al. (1988) provide a number of checklists each dealing with different aspect of human factors and human resources concerned with information technology, including: working conditions; equipment design; usability; job quality; organisational effectiveness. Checklists such as these may become dated as new approaches are devised but their essence stays the same. Examples such as these are useful in providing models for developing checklists tailored to specific needs.

12 *Verbal protocol analysis* is concerned with obtaining a verbal account of the operator's thoughts, considerations and motivations when carrying out a task. There is a problem that asking people to speak aloud, when they would not normally do

so, may invalidate what they have to say. Talking is a secondary task that can interfere with the main task being carried out. Furthermore, what operators say, may not actually reflect their reasons for performing as they do. Papers concerned with verbal protocol analysis vary according to the extent to which they claim that the verbal protocols collected represent mental events or merely provide the analyst with useful insight about the task. Practical accounts of verbal protocol analysis are given in Bainbridge and Sanderson (1995) and Kirwan and Ainsworth (1992).

13 An analyst can use many (or few) methods of task analysis during the course of an HTA. Kirwan and Ainsworth (1992) provide a useful source. The book provides a brief outline of 41 methods, and provides detailed accounts and illustrative case studies of several of the more common of these. These include illustrations of *link-analysis*, which enables the analyst to understand how different areas of the task, different instruments, items of equipment and different personnel are linked to one another through the execution of the task. This has implications for workplace layout and job-design. *Timeline analysis* examines the temporal interdependency of actions within tasks. This can show how actions follow sequences in response to events and which actions need to be undertaken at the same time. This has implications for operator workload and allocation of tasks within teams.

Stanton and Young (1999) provide a practical guide to several useful methods that can be used in task analysis projects which includes many practical examples. Accounts are given of methods for analysing computer interfaces, workplace layout, human error, as well as practical accounts of link analysis, observation, questionnaires, repertory grids and interviews.

Diaper (1989c) provides a collection of methods that have been developed to support HCI projects. These include *Task Action Grammar* (TAG) which provides a formal method for relating actions to tasks in the human computer interaction domain, with respect to computer program design in particular (Payne and Green, 1989), methods for the formal identification of task knowledge (Johnson and Diaper) and methods for managing system development (Walsh).

Monk and Gilbert's (1995) collection of chapters includes accounts of some of the techniques covered in Diaper's collection, but also includes accounts of other approaches to examining work, including *ethnography* (Blomberg, 1995), *conversational analysis* (Greatbatch et al. 1995), *activity theory* (Blackler, 1995) and *organisational analysis* (Jones). These tend to be self-contained approaches which distance themselves from task analysis approaches - their authors would probably regard the more mainstream approaches of human factors as too focused on prescribed goals without properly appreciating how context influences the real motivations and behaviours of people at work. This is not a wholly fair criticism. All analysts can exercise sensitivity to other influences, such as the affects of a worker's individual motivations and organisational dynamics. A task analyst should adopt a broad perspective when examining tasks, and such breadth is encouraged by reading about other approaches.

Task analysts need to be equipped with a number of techniques, some of which may not be seen as task analysis methods. Robson (1993) provides a very practical account of many enquiry methods, including designing small surveys, carrying out evaluations, making observations, interviewing and questionnaire design.

14 One of the most useful and straightforward methods for obtaining information about how operators use information to make decisions is the *withheld information technique*. This has been used in a number of studies into fault-finding (e.g. Marshall et al., 1981), but it has not been widely reported. It is often useful when investigating issues concerned with the strategic use of information by operators within a practical HTA project. A brief account of it will be given here for future reference.

The Withheld information technique

There are three main stages to consider with regard to using the withheld information technique:

1. Preparing materials for the investigation.

2. Collecting information from the operator.

3. Using the results.

1.   Preparing materials for the investigation.

  i.   List a range of problems or events with which the operator might be expected to cope. This can include common problems, unusual problems and composite problems. It could include the actions that could be taken to deal with the problem.

  ii. List the information that is typically available for the operator to use. This can include information that is readily available in the workplace and information that might be obtained by having to go elsewhere or by contacting other personnel. It is possible to try out different information sets to decide whether additional information would be helpful or removing certain information would be detrimental.

  iii. Prepare a matrix where the problems or events (from i.) are each assigned to a row and the information sources (from ii.) are each assigned to a column.

  iv. Using task experts, consider each row in turn and predict the information that would prevail if that problem or event were to occur. Enter this in the respective cell. If a range of possible information values can occur, then indicate this in the cell. In fault-finding exercises, a matrix completed in the way is called a fault-symptom matrix.

2. *Collecting information from the operator.*

   i. Operators or trainees are then told that they will be presented with a problem which they must try to diagnose. They may be provided with the set of information, from 1ii. above, or they may be allowed to consider any information they choose.

   ii. The analyst then, covertly, selects one of the problems or events.

   iii. Operators are told that there is a problem with the system. They are then required to select the first item of information whose value they wish to obtain.

   iv. When the operator asks for a specific item of information, the analyst specifies the value that is written in the respective cell for the event selected earlier (in 2ii.).

   v. The operator then considers this information, then either requests another item of information or, if the operator thinks the problem has been resolved, he or she will then state what this problem or event is.

   vi. Sometimes the operator will declare a solution before eliminating other possibilities. The analyst may, therefore, choose to inform the operator of this, so that the operator can explore other information.

3. *Using the results.*

   This sort of exercise can be used in different ways. It can show where the situation contains insufficient information to support the ways in which people want to use information. It can show the sorts of strategy that people (experts) use when carrying out diagnosis, and so this can be helpful in modelling expert performance. It can show whether people are likely to make errors when dealing with certain types of problem. It can reflect on operator skills (and can, therefore, be used in training programmes) by showing whether people diagnose prematurely with insufficient information, whether they collect more information than they require or whether they ask questions that do not enable them to reduce uncertainty.

   The withheld exercise is artificial in that it does not represent how information is actually *presented*. Therefore, it will not support pattern recognition strategies. But it is a quick way of obtaining some useful insights into how people deal with different situations.

15 Hierarchical organisation is a common feature in task analysis methods. Many approaches have chosen to break down complex activity into smaller components which may then be better understood, for example, FAST (Functional Analysis System Technique, see Creasy, 1980) which exploits the idea that the functioning of systems is dependent on the functioning of their subsystems.

Different principles may be adopted to govern the nature of this decomposition. Some approaches seek to describe tasks in terms of a set of standard elements, while some other approaches seek to decompose tasks in terms of a standard set of levels of redescription.

Earlier approaches to scientific management included the work by Gilbreth in which tasks were broken down into standard elements of movement (see Blum and Naylor, 1968, for example or most other texts on industrial psychology). In a similar fashion, Crossman's (1956) 'sensorimotor chart' approach sought to describe tasks by considering the contribution of standard components - through each limb, the eyes and the brain. Skill emerged as these elements were carried out effectively and coordinated properly by the brain. These ideas were applied to the analysis of manual skills in industry through the work of Seymour (1966). Another strategy for decomposition, which resulted in identifying standard task elements, was the approach of Fleischman (e.g. Fleischman and Quaintance, 1984). In numerous studies, Fleischman carried out *factor analysis* on different sets of test scores in order to identify the underlying factors (skills) governing the execution of more complex tasks. Thus, the skills in carrying out complex tasks could be decomposed in terms of a standard set of skills. HTA does not adopt a standard set of components for redescription, apart from distinguishing between operations and plans.

Strategies for decomposition into fixed levels include *GOMS* (Goals, Operators, Methods and Selection rules, Card et al., 1983). *Goals* are a person's intentions, *operators* are the individual things the person will do, *methods* are the ways in which basic operations are combined to meet goals, and *selection rules* are the means by which the person chooses between different methods in order to achieve the goal. Thus, the analyst will represent the task in terms of four major levels of description. The authors of HTA deliberately avoided adopting any fixed level of analysis, because this would mean that some parts of a task would be developed in excessive detail while other parts may not be described in sufficient detail. It is worth recognising that GOMS tends to be applied in contexts where a uniform level of detail may be appropriate, whereas this would be problematical when dealing with more complex tasks in other domains.

16 The *critical incident technique* (Flanagan, 1954) focuses on incidents that have occurred - accidents or near misses. The analyst seeks to identify the factors that have caused the problem. This can be a useful starting point for HTA in the knowledge that it directs attention to aspects of the workplace which have proven to be unsatisfactory.

17 *Open System Task Analysis* (Eason, 1988) is a method which uses the concepts of system thinking to consider the functions of the work environment of interest then lists its *inputs*, its *outputs* and the *functional transformations* within the work

environment. In this way the areas of work activity likely to be of particular interest are identified.

Other systems for providing a richer assessment of an intervention include *Soft Systems Analysis* and *Organisational Analysis*. *Soft Systems Analysis* (Checkland, 1981) was developed from systems analysis in order to deal with issues of social concern. The approach advocates a methodology whereby the analyst considers a problem, provides a root definition of the purpose of the system within which the problem resides, then models this system using various analysis tools, then compares this model with the real world with a view to identifying solutions to problems.

Organisational Analysis (see Dawson, 1986; Jones, 1995) recognises the interaction between various elements within an organisation - technology, tasks and methods, organisational structures and the environment. The organisational analyst seeks to understand how problems may emerge or may be resolved by a proper appreciation of the dynamics within the organisation.

Each of these methods is designed to focus where effort is best directed. Unless the task analyst focuses effectively, there is a risk that effort will be misdirected. However, task analysis is time consuming and it would be inappropriate to insist that one of these methods is first carried out fully before the task analysis is undertaken. But some experience of these methods is helpful because it alerts the analyst to how tasks being examined can be influenced.

18 Shepherd and Hinde (1989) developed a mapping of the interactions between different tasks within organisations to show how problems in one area are often referred from other areas - thus the analyst's attention may be diverted from the source of a problem by the symptoms that are presented.

19 *Cost-benefit analysis* is essentially concerned with making choices and/or justifying courses of action where substantial investment is involved in order to achieve an unpredictable return. In a commercial context a company would wish to see a profit from the investment, thus, the amount of cash returned on the investment must exceed the amount of cash invested. Making such decisions seems reasonably straightforward. In practice, it is far from straightforward, since anticipating a full range of potential cost items is complicated and estimating what those costs will be with any confidence is tricky. Some costs, such as the costs of hardware, will be fixed and negotiable. Other costs, such as the costs of maintenance or marketing will vary in accordance with events which are difficult to anticipate. Benefits are also speculative. Cost-benefit analysis is simple in its conception but extremely complex and often intractable in its execution. Levin (1983) deals with a number of these issues. Rouse and Boff (1997) discuss the wider issues concerned with the cost-benefits of human factors.

20 In principle, an analyst could embark on the HTA framework, then make a decision almost immediately to focus analysis on one area using a different task analysis method. All the framework will have done is focus the analyst's attention on an area of criticality.

Chapter 3: Plans and complexity

Chapter 4: Flexibility, constraint, cognition and context

21 Cognitive task analysis methods include methods for data collection and methods for modelling and representing cognition. Reference is made in Notes 13 and 15 to a number of *cognitive task analysis* methods concerned with data collection. Eberts (1997) provides a comprehensive review of methods of cognitive modelling and representation.

22 Variations in operator strategy as a consequence of experience and context are accounted for in models such as Rasmussen's *skill-rule-knowledge* approach (1980, 1986) - see Note 10 - and Reason's *GEMS* (Generic Error Modelling System: Reason, 1990). Reason's approach is based on Rasmussen's and shows how when tasks are carried out by people with different levels of expertise, performance is affected in different ways because the operator attempts to deal with the task in different ways.

23 A more detailed account of this task is given in Williams (1992).

24 *Signal-detection theory* is a well-known psychological model which has helped psychologist to understand the nature of detecting signals in situations of uncertainty. In essence, people are forced to take risks in making such judgements and the strategy for taking these risks is influenced by factors such as expectation and outcome rather than simply a judgement of the magnitude of the signal. Simple accounts are given in Wickens (1992) and Wickens and Carswell (1997). My reason for including a reference to signal-detection theory is not to emphasise that theory, but to point out that many such theories and models from experimental psychology can be manifested in elements of real tasks observed whilst carrying out HTA. The analyst can rarely translate findings from experimental psychology directly to real tasks, but knowledge of some of these models is often helpful when trying to make sense of certain tasks. Texts such as Wickens (1992) are ideal for providing such background.

Chapter 5: Representing and recording HTA

25 The consequences of error in hazardous systems such as process control, transportation and medical situations can be extremely serious. A number of methods have been devised to enable systematic examination of tasks and processes with respect to the possible occurrence of human error in these situations. Often, potentially hazardous installations or situations, such as nuclear plant, railway systems or public venues need to demonstrate to safety inspectorates that the systems proposed are safe, that they are properly constructed and appropriately staffed, operated and managed. This has prompted the development of a number of *Human Error Analysis* (HEA) methodologies whose purpose is to conduct a systematic evaluation to identify potential problems and assess their risk. These various methods are described, reviewed and illustrated extensively in Kirwan (1990, 1994),

Human error analysis methods first require the analyst to describe the tasks, in order that the steps entailed can be subjected to scrutiny. Scrutiny is usually carried out using one of several error classifications or taxonomies. One example is Swain and Guttman's (1983) use of external error modes, where the analyst makes judgements about the consequences of actions on carrying out the process - actions could be omitted, carried out too early or too late, and so on. A different approach is SHERPA (Systematic Human Error Reduction and Prediction Approach: e.g. Embrey (1986) which uses a taxonomy based on GEMS (see note 22). This uses a more elaborate psychological classification of the ways in which errors are likely to be committed at various stages in the task - for example, *stereotype takeover*, *mistake using alternative, slip of memory*. See Kirwan (1990) for a full list.

Each of these approaches clearly depends upon a proper analysis of the task and the application of a suitable classification scheme. HTA has served effectively as part of many HEA projects because it sets out the various steps of the task. To apply any human error classification to a task analysis usually requires that a tabular format is adopted where columns to the left contain the task steps — from the task analysis - and further columns are added in which the analyst can record whether or not the step could invoke that type of error. Table 5.5 illustrates this with a simple example.

26 The reader interested in software to support HTA should look at Bass et al. (1995). This describes a promising prototype for an integrated HTA tool in which text could be entered in table or diagram mode and provides effective output facilities.

27 Requests for software to speed-up the processes of HTA (and task analysis generally) are commonplace. In preparing diagrams and tables for this book, I have generally used 3 computer packages - *Microsoft Word*™, *Aldus Intellidraw*™ and *Microsoft Excel*™. Microsoft Word is useful because of the outline facility (see Figure 5.6).

This enables the analyst to move text to different levels and then number the result in a consistent manner. Version 5.1 of Word on the Apple Macintosh (equivalent to version 3.1 on Microsoft Windows™) is particularly useful because of the control given to the analyst in numbering. In later versions of Word, the outline facility is more automated and seems to have a mind of its own, with numbers of operations changing when it is not always appropriate, especially when the analyst wishes to set the sequence out in accordance with Figure 5.6. When the analysis has been set out in Word, it can then be translated into a diagram using Intellidraw. This is a clever drawing package which gives the user considerable control over boxes and joining lines and enables text flow between linked boxes. Unfortunately, the package is no longer produced and these functions tend not to be provided, so easily, in more recent software. There are a number of packages for producing hierarchical diagrams from text to produce organisation charts, but these offer limited layout facilities. This is unsuitable if the analyst wishes to layout as much of the diagram on a single page, but may be satisfactory if multiple pages are used with a limited number of boxes on each. The Word file can also be inserted into Excel to produce tables with ease. Later versions of Excel are particularly effective in providing many ways of formatting boxes, including text wrapround within cells. I am not making any strong claims for the software I have used, because people familiar with other packages may find equally useful ways of achieving the same end.

Chapter 6: Analysis of tasks - some illustrations

28 Process control has received considerable attention in the applied research literature. See Edwards and Lees (1974) for an early collection of papers, many of which contain fundamental arguments which have laid the foundations for later work. A more recent review is provided by Moray (1997). Bainbridge (1987) considers several of the unexpected outcomes for human control of automated systems in this and other areas.

29 Human factors research in Air-Traffic Control is reviewed by Hopkin (1995).

·30 This example is developed more fully in Shepherd (1992).

31 Shepherd and Hinde (1989) present a variant on HTA to describe tasks and functions in organisations generally, showing how the same functions apply in different contexts irrespective of the size of the organisation.

Chapter 7: Making human factors design decisions within HTA

32 See Howarth (1995) for a review of visual aspects of ergonomics and Haslegrave (1995) for a review of the auditory environment.

33 See Pheasant (1995) for a review of anthropometry and workspace design. Corlett and Clark (1995) provide a broad and practical account of all issues in workplace ergonomics.

34 There are many texts concerned with aspects of designing for computer interfaces. Useful reviews are provided by Lansdale and Ormerod (1994), Wilson and Rajan (1995) and Preece et al. (1994).

35 Similar dimensions are discussed in Meister (1985).

36 The concept of *affordance* is discussed by Norman (1988) and Lansdale and Ormerod (1994).

37 The concept of the system life cycle and its relationship to human factors decisions is discussed in Chapter 1 of Kirwan and Ainsworth (1992).

Chapter 8: Teams and jobs

38 Medsker and Campion (1997) review issues of team design. Pennington et al. (1992) describe an exercise in which HTA is used to assess staffing needs in a control room.

39 A detailed case study showing the application of timelines to workload assessment is given in Pennington et al. (1992). McLeod and Sherwood-Jones (1992) demonstrate how timings of operations can be used to predict operator workload.

40 The mere coincidence of tasks may not be sufficient to decide whether they can coexist or whether they interfere with one another. They may be more or less mentally or physically demanding. Meshkati et al. (1995) discuss issues concerned with mental workload. Issues associated with dynamic work are discussed by Kilbom (1995) and those concerned with the effects of static posture in work by Corlett (1995).

Chapter 9: Information and skill

41 See Pheasant (1995) for a review of anthropometry and workspace design. Environmental ergonomics refers to the visual environment (see Howarth, 1995), auditory and noise environment (see Haslegrave, 1995) and the thermal environment (see Parsons, 1995).

42 Information requirements specification is a topic dealt with by Astley and Stammers (1988), Shepherd, (1993) and Ormerod et al. (1998).

43 The world 'planning' used in a general operating sense relates to the word 'plan' as used in HTA. In HTA, a plan is a statement of the conditions under which operations are carried out in order to meet a goal. Plans in HTA state times, sequences and events that cue subsequent actions.

44 There are several texts concerned with providing information displays and controls. Bullinger et al. (1997) deal with visual displays, while Bennett et al. (1997) deal with issues concerned with the design of controls. Corlett and Clark (1995) offer a practical account of equipment and workplace design. There are many texts concerning design for human computer interaction in which the issue of representation and control is central (e.g. Preece et al. 1994; Lansdale and Ormerod, 1994; Baber, 1997) In all of these topics there are concerns with both the *physical* and *physiological* aspects of interaction - is the design of displays suited to the operators visual system and can controls be manipulated appropriately by the operator? There are also *psychological* concerns - do displays represent information in a way that can be readily interpreted by the operator and are controls designed so that they work in accordance with the operator's expectation and provide suitable feedback to regulate behaviour.

Chapter 10: HTA and training

45 HTA has often been associated with training design. This was primarily because the original authors were interested in this application and frequently wrote about training issues (e.g. Annett and Duncan, 1967; Annett, 1991; Duncan, 1974; Duncan and Gray, 1975). But it was also their concern that task analysis should be neutral with respect to the solution adopted - the task analysis showed where performance problems resided and which solutions should be favoured. That their main interests in developing solutions rested with training, was coincidental.

The chapter only addresses practical issues concerning HTA and training. It does not deal with training issues in general. Swezey and Llaneras (1997) and Patrick (1992) provides a comprehensive review of training research.

46 This model is developed in Shepherd and Kontogiannis (1998).

47 There are many accounts of managing the processes of training design and development. A common approach is the *Instructional Systems Development* (ISD) model (Branson et al., 1975). Comprehensive accounts are given in Swezey and Llaneras (1997) and Patrick (1992).

48 *Part-task training* is an issue which has received considerable theoretical attention within training, research. Because it is often difficult to generalise from laboratory studies to real situations, the trainer should use more discretion in applying these ideas. Therefore, it is important that practitioners understand what is happening to their trainees and avoid resorting to dogma. Lintern (1991) and Patrick (1992) provide useful accounts of these issues.

49 The psychological issues concerned with *simulation design* are discussed by Patrick (1992).

Chapter 11: Designing support documentation

50 Much of the practical advice to guide the design of documentation comes from the psychology and ergonomics of language. There are issues concerning the presentation of text and the nature of the language used. The basic issues of written instructions are contained within a classic paper by Broadbent (1977). Useful references to guide the design of documentation are Hartley (1985, 1995) and Wright (1987).

Chapter 12: Some human resource management issues

51 Human resource management is a broad topic dealing with issues that extend far beyond the limited coverage of this chapter. There are numerous texts dealing with these issues, for example, Attwood (1996).

References

Annett, J. (1991) Chapter 2 - Skill acquisition. In Morrison, J. (ed) *Training for Performance*. Chichester: John Wiley & Sons Ltd, pp. 13-51.

Annett, J. and Duncan. K. D. (1967) Task Analysis and Training Design. *Occupational Psychology, 41*, pp. 211-21.

Annett, J., Duncan, K. D., Stammers R. B. and Gray. M. J. (1971) *Task Analysis*. London: HMSO.

Astley, J. A. and R. B. Stammers. (1988) Adapting hierarchical task analysis for user-system interface design. *Contemporary Ergonomics*.

Attwood, M. (1996) *Personnel Management* 3rd edition. Basingstoke: Macmillan.

Baber, C. (1997) *Beyond the Desktop*. London: Academic Press.

Bainbridge, L. (1987) Ironies of automation. In Rasmussen, J., Duncan, K. D. and Leplat, J. (eds) *New technology and human error*. Chichester: John Wiley & Sons, pp. 271-83.

Bainbridge, L. and Sanderson, P. (1995) Verbal protocol analysis. In Wilson, J. R. and Corlett, E. N. (eds) *Evaluation of Human Work* (2nd ed). London: Taylor and Francis, pp. 169-201.

Bass, A., Aspinall, J., Walters, G. and Stanton, N. (1995) A software toolkit for hierarchical task analysis. *Applied Ergonomics, 26*, pp. 147-51.

Bennett, K. B., Nagy, A. L. and Flack. J. M. (1997) Visual displays. In Salvendy, G, (ed) *Handbook of Human Factors and Ergonomics*. New York: Wiley-Interscience, pp. 659-96.

Blackler, F. (1995) Activity Theory, CSCW and Organisations. In Monk, A. F. and Gilbert, N. (eds) *Perspectives on HCI: Diverse Approaches*. London: Academic Press, pp. 223-48.

Blomberg, J. L. (1995) Ethnography: aligning field studies of work and system design. In Monk, A. F. and Gilbert, N. (eds) *Perspectives on HCI: Diverse Approaches*. London: Academic Press, pp. 175-97.

Blum, M. and Naylor, J. (1968) Industrial Psychology - Theoretical and Social Foundations. New York: Harper & Row.

Branson, R. K., Rayner, G. T. Coxx, J. L., Furman, J. P., King, F. J. and Hannum, W. J. (1975) *Interservice Procedures for Instructional System Development: Executive Summary and Model*. Fort Benning, GA: U.S. Army Combat Arms Training Board.

Broadbent, D. (1977) Language and ergonomics. *Applied Ergonomics, 8*, 15-8.

Bullinger, H.-J., Kern, P. and Braun, M. (1997) Controls. In Salvendy, G, (ed) *Handbook of Human Factors and Ergonomics*. New York: Wiley-Interscience, pp. 697-728.

Card, S. K., T. P. Moran and A. Newell. (1983) *The Psychology of Human-Computer Interaction*. Hillsdale, NJ: Lawrence Erlbaum Associates.

Checkland, P. B. (1981) *Systems Thinking, Systems Practice*. Chichester: Wiley.

Clegg, C., Warr, P., Green, T., Monk, A., Kemp, N., Allison, G. and Lansdale, M. (1988) *People and Computers: How to Evaluate your Company's New Technology*. Chichester: Ellis Horwood Ltd.

Corlett, E. N. (1995) The evaluation of posture and its effects. In Wilson, J. R. and Corlett, E. N. (eds) *Evaluation of Human Work* (2nd ed). London: Taylor and Francis, pp. 662-714.

Corlett, E. N. and Clark, T. S. (1995) The Ergonomics of Workspaces and Machines. London: Taylor and Francis, 1995.

Creasy, R. (1980) Problem solving the FAST way. *Proc. Society of American Value Engineers Conference*. Irving, Texas, USA, pp. 173-5.

Crossman, E. R. F. W. (1956) Perceptual activities in manual work. *Research, 9*, pp. 42-9.

Dawson, S. (1986) *Analysing Organisations*. London: MacMillan Education.

Diaper, D. (1989a) *Task Analysis for Human-Computer Interaction*. Chichester: John Wiley & Sons Ltd.

Diaper, D. (1989b) Task analysis for knowledge description (TAKD); the method and an example. In Diaper. D. (ed) *Task Analysis for Human-Computer Interaction*. Chichester: Ellis Horwood, pp. 210-37.

Diaper, D. (1989c) Task observation for human-computer interaction. In Diaper. D. (ed) *Task Analysis for Human-Computer Interaction*. Chichester: Ellis Horwood, pp. 210-37.

Duncan, D. K. (1974) Analytical Techniques in Training Design. In Edwards, E. and Leeds, F. P. (eds) *The Human Operator and Process Control*. London: Taylor and Francis, pp. 283-320.

Duncan, K. and M. Gray. (1975) Functional Context Training: A Review and an Application to a Refinery Control Task. *Le Travail Humain, 38*, pp. 81-95.

Eason, K. D. (1988) *Information Technology and Organisational Change*. London: Taylor and Francis.

Eberts, R. (1997) Cognitive modelling. In Salvendy, G, (ed) *Handbook of Human Factors and Ergonomics*. New York: Wiley-Interscience, pp. 1328-74.

Edwards, E. and F. P. Lees. (1974) *The Human Operator in Process Control*. London: Taylor and Francis.

Embrey, D. E. (1986) SHERPA: a systematic human error reduction and prediction approach. *International Topical Meeting on Advances in Human Factors in Nuclear Power Systems*. Noxville, Tennessee.

Flanagan, J. C. (1954) The critical incident technique. *Psychological Bulletin, 51*, pp. 327-58.

Fleischman, E. A. and Quaintance, M. K. (1984) *Taxonomies of Human Performance: The Description of Human Tasks*. Orlando, Fl: Academic Press.

Gagne, R. M. The Conditions of Learning. London: Holt, Reinhart and Winston, 1970.

Gagne, R. M., Briggs, L. J. and Wager, W. W. (1988) *Principles of Instructional Design*. New York: Holt, Reinhart and Winston, Inc.

Greatbatch, D., Heath, C., Luff, P. and Campion, P. (1995) Conversation analysis: human -computer interaction and the general practice consultation. In Monk, A. F. and Gilbert, N. (eds) *Perspectives on HCI: Diverse Approaches*. London: Academic Press, pp. 199-222.

Hartley, J. (1985) *Designing Instructional Text*. New York: Nichols.

Hartley, J. (1995) Is this chapter any use? Methods for evaluating text. In Wilson, J. R. and Corlett, E. N. (eds) *Evaluation of Human Work* (2nd ed). London: Taylor and Francis, pp. 283-309.

Haslegrave, C. M. (1995) Auditory environment and noise assessment. In Wilson, J. R. and Corlett, E. N. (eds) *Evaluation of Human Work* (2nd ed). London: Taylor and Francis, pp. 506-40.

Hopkin, D. V. (1995) *Human Factors in Air Traffic Control*. London: Taylor and Francis.

Howarth, P. (1995) Assessment of the visual environment. In Wilson, J. R. and Corlett, E. N. (eds) *Evaluation of Human Work* (2nd ed). London: Taylor and Francis, pp. 445-82.

Jenkins, G. M. (1969) The systems approach. *Journal of Systems Engineering, 1*.

Johnson, P. (1989) Supporting system design by analysing current task knowledge. In Diaper. D. (ed) *Task Analysis for Human-Computer Interaction*. Chichester: Ellis Horwood, pp. 160-85.

Jones, M. (1995) Organisational Analysis and HCI. In Monk, A.F. and Gilbert, N. (eds) *Perspectives on HCI: Diverse Approaches*. London: Academic Press, pp. 249-69.

Kilbom, A. (1995) Measurement and assessment of dynamic work. In Wilson, J. R. and Corlett, E. N. (eds) *Evaluation of Human Work* (2nd ed). London: Taylor and Francis, pp. 640-61.

Kirwan, B. (1994) *A Guide to Practical Human Reliability Assessment*. London: Taylor & Francis.

Kirwan, B. (1995) Human reliability assessment. In Wilson, J. R. and Corlett, E. N. (eds) *Evaluation of Human Work* (2nd ed). London: Taylor and Francis, pp. 921-68.

Kirwan, B. and Ainsworth, L. K. (eds) (1992) *The Task Analysis Guide*. London: Taylor and Francis.

Lansdale, M. W. and Ormerod, T. C. (1994) *Understanding Interfaces: A Handbook of Human-Computer Dialogue*. London: Academic Press.

Levin, H. M. (1983) *Cost-Effectiveness: A Primer*. Beverly Hills: Sage.

Lintern, G. (1991) Chapter 6 - Instructional strategies. In Morrison, J. (ed) *Training for Performance*. Chichester: John Wiley & Sons Ltd, pp. 167-91.

Marshall, E. C., Duncan, K. D. and Baker, S. (1981) The role of withheld information in the training of process plant fault diagnosis. *Ergonomics, 24*, pp. 711-24.

McLeod, R. W. and Sherwood-Jones, B. M. (1992) Simulation to predict operator workload in a command system. In Kirwan, B. and Ainsworth, L. K. (eds) *A Guide to Task Analysis*. London: Taylor and Francis, pp. 301-10.

Medsker, G. J. and Campion, M. A. (1997) Job and team design. In Salvendy, G, (ed) *Handbook of Human Factors and Ergonomics*. New York: Wiley-Interscience, pp. 450-89.

Meister, D. (1985) *Behavioural Foundations of System Development*. Malabar, Florida: Robert E. Krieger Publishing Company, Inc.

Meshkati, N., Hancock, P. A., Rahimi, M. and Dawes, S. M. (1995) Techniques in mental workload assessment. In Wilson, J. R. and Corlett, E. N. (eds) *Evaluation of Human Work* (2nd ed). London: Taylor and Francis, pp. 749-82.

Miller, G. A., Gallanter, E. and Pribram, K. (1960) *Plans and the Structure of Behaviour*. New York: Holt, Reinhart and Winston.

Miller, R. B. (1953) *A method for man-machine task analysis*. Wright Air Development Center, Wright-Patterson AFB, Ohio.

Miller, R. B. (1966) Task taxonomy: science or technology. In Singleton, T., Easterby, R. S. and Whitfield, D. C. (eds) *The Human Operator in Complex Systems*. London: Taylor and Francis.

Monk, A. and Gilbert, N. (eds). (1995) *Perspectives on HCI: Diverse Approaches*. London: Academic Press.

Moray, N. (1981) Feedback and the Control of Skilled Behaviour. In Holding, D. (ed) *Human Skills*. John Wiley & Sons Ltd, pp. 15 - 39.

Moray, N. (1997) Human factors in process control. In Salvendy, G, (ed) *Handbook of Human Factors and Ergonomics*. New York: Wiley-Interscience, pp. 1944-72.

Norman, D. A. (1988) *The Psychology of Everyday Things*. New York: Basic Books.

Ormerod, T. C., Richardson, J. and Shepherd, A. (1998) Enhancing the Usability of a Task Analysis Method: A Notation and Environment for Requirements Specification. *Ergonomics, 41*, 1642-63.

Parsons, K. C. (1995) Ergonomics assessment of thermal environments. In Wilson, J. R. and Corlett, E. N. (eds) *Evaluation of Human Work* (2nd ed). London: Taylor and Francis, pp. 483-506.

Patrick, J. (1992) *Training Research and Practice*. London: Academic Press - Harcourt Brace Jovanovich, Publishers.

Payne, S. J. and Green, T. R. G. (1989) Task-action grammar, the model and its developments. In Diaper. D. (ed) *Task Analysis for Human-Computer Interaction*. Chichester: Ellis Horwood, pp. 75-107

Pennington, J., Joy, M. and Kirwan, B. (1992) A staffing assessment for a local control room. In Kirwan, B. and Ainsworth, L. K. (eds) *A Guide to Task Analysis*. London: Taylor and Francis, pp. 289-99.

Pheasant, S. T. (1995) Anthropometry and the design of workspaces. In Wilson, J. R. and Corlett, E. N. (eds) *Evaluation of Human Work* (2nd ed). London: Taylor and Francis, pp. 557-73.

Preece, J., Rogers, Y., Sharp, H., Benyon, D., Holland, S. and Carey, T. (1994) *Human-Computer Interaction*. Wokingham: Addison-Wesley.

Rasmussen, J. (1980) The Human as a Systems Component. In Smith, H. and T. Green, T. (eds) *Human Interaction with Computers*. London: Academic Press, p. 67 - 96.

Rasmussen, J. (1986) *Information Processing and Human-Machine Interaction: An Approach to Cognitive Engineering*. North Holland.

Reason, J. T. (1990) *Human Error*. Cambridge: Cambridge University Press.

Robson, C. (1993) *Real World Research*. Oxford: Blackwells.

Rouse, W. B. and Boff, K. B. (1997) Assessing cost/benefits of human factors. In Salvendy, G, (ed) *Handbook of Human Factors and Ergonomics*. New York: Wiley-Interscience, pp. 1617-34.

Salvendy, G (ed.) (1997) *Handbook of Human Factors and Ergonomics*. New York: Wiley-Interscience.

Seymour, W. D. (1966) *Industrial Training for Manual Operation*. London: Pitman.

Shepherd, A. (1993) An approach to information requirements specification for process control tasks. *Ergonomics, 36*, 805-17.

Shepherd, A. (1985) Hierarchical task analysis and training decisions. *Programmed Learning and Educational Technology, 22*, pp. 162-76.

Shepherd, A. (1992) Maintenance Training. In Kirwan, B. and Ainsworth, L. K. (eds) *A Guide to Task Analysis*. London: Taylor and Francis, pp. 327-39.

Shepherd, A. (1995) Task analysis. In Monk, A.F. and Gilbert, N. (eds) *Perspectives on HCI: Diverse Approaches*. London: Academic Press, pp. 145-74.

Shepherd, A. and Hinde, C. J. (1989) Mimicking the training expert: a basis for automating training needs analysis. In Bainbridge, L. and Quintanilla, S. A. R. (eds) *Developing Skills with Information Technology*. Chichester: Wiley, pp. 153-76.

Shepherd, A. and Kontogiannis, T. (1998) Strategic task performance: a model to facilitate the design of instruction. *International Journal of Cognitive Ergonomics, 2*, pp. 349-72.

Stammers, R. B., Carey, M. S. and Astley, J. A. (1991) Task Analysis. In Wilson, J. R. and Corlett, E. N. (eds) *Evaluation of Human Work* (1st ed). London: Taylor and Francis.

Stanton, N. and Young, M. S. (1999) *A Guide to Methodology in Ergonomics*. London: Taylor and Francis.

Swain, A. D. and Guttmann, H.E. (1983) A Handbook of Reliability Analysis with Emphasis on Nuclear Power Plant Applications. Nureg/CR-1278 (Washington, DC: USNRC).

Swezey, R. W. and Llaneras, R. E. (1997) Models in training and instruction. In Salvendy, G, (ed) *Handbook of Human Factors and Ergonomics*. New York: Wiley-Interscience, pp. 514-77.

Walsh, P. Analysis for task object modelling (ATOM). In Diaper. D. (ed) *Task Analysis for Human-Computer Interaction*. Chichester: Ellis Horwood, pp. 186-209.

Wickens, C. (1992) *Engineering Psychology and Human Performance*. New York: Harper Collins.

Wickens, C. D. and Carswell, C. M. (1997) Information processing. In Salvendy, G, (ed) *Handbook of Human Factors and Ergonomics*. New York: Wiley-Interscience, pp. 109-29.

Williams, J. C. (1992) A method for quantifying ultrasonic inspection effectiveness. In Kirwan, B. and Ainsworth, L. K. (eds) *A Guide to Task Analysis*. London: Taylor and Francis, pp. 341-53.

Wilson, J. R. and E. N. Corlett. (eds.) (1995) *Evaluation of Human Work*. London: Taylor and Francis.

Wilson, J. R. and Rajan, J. A. (1995) Human machine interfaces for systems control. In Wilson, J. R. and Corlett, E. N. (eds) *Evaluation of Human Work* (2nd ed). London: Taylor and Francis, pp. 357-405.

Wright, P. (1987) Issues of content and presentation in document design. In: *Handbook of Human-Computer Interaction*, edited by M. Helander. Amsterdam: North-Holland.

Index

A

access to help 149
action feedback 14
actions 187
activating or deactivation equipment 187
activity theory 249
adjusting equipment 187
affordance 257
air-traffic control [task] 111, 256
anthropometry and workspace design
    257, 258
appraising and developing staff 238
assessment 204
auditory environment 257
automatic control 146

B

batch processing task 105, 106, 143, 178,
    180
behaviour 239
boiling an egg task 44, 177
breaking the information and control loop
    180

C

changing a cartridge in a laser printer task
    104
checklists 30, 143, 219
chlorine balancing task 57, 178
choice and decision-making 47
client 21
coaching and instructing 172
cognition and flexibility 243
cognitive coordination 50
cognitive modelling 254
cognitive task analysis 3, 69, 239, 254
coincidence 81
collaborating on a common goal 159
collaboration between nurses and doctors
    [task] 162
columns in a task analysis table 93

communications 187
complex plans 56
composite plans 56
computer aids in recording task analysis 98
computer-based control system [task] 182
computer-based accountancy [task] 74
concurrent operations 49
context 42, 243
context and constraint and design decisions
    147
contingencies 164
contingent sequence 44
continuous process control [task] 105, 109
control feedback 14
controllability of events 147
controlling complex systems [task] 44
controlling the actions of colleagues [task]
    173
controls 4, 176
conversational analysis 249
coordination 49
cost-benefit analysis 27, 38, 152, 253
costs of training support 149
criteria 21
critical incident technique 34, 252
criticality 23
cumulative part-task training 212
customer service [task] 116
cycles 52

D

decision aids 221
decision outcome 84
decision trees 222
degraded feedback 179
delegating problems 168
delivering training programmes 208
design choices 143, 151
design constraints 20, 37
design methodologies 143
design options 72, 136, 137
detailed design 136, 152

detecting flaws in welds [task] 78
diagnosis 190
difficulty of the task 147
discrimination 72
dispatch preparation [task] 75
distillation 51, 139, 178
documentation 259

E

effectiveness of training 208
engaging in a task 199
engineering and technological issues 153
environmental and movement constraints
      149, 153
environmental ergonomics 140, 258
ergonomics checklists 248
error classifications 255
ethnography 249
exchanging task information 173

F

FAST (Functional Analysis System Tech-
      nique 251
feedback and control in performance 11
fidelity in simulation 214
fixed-sequence plans 43
flexibility 68
flow charts and rule-sets 97, 222
focus and bias 34, 71, 74
formal training programme 205
frequency of demand 81
functional design 141

G

GEMS (Generic Error Modelling System)
      254, 255
giving instruction 188
goal 1
goal context 78
goal directed behaviour 11
GOMS (Goals, Operators, Methods and
      Selection) 252
guidelines 143

H

hierarchical diagrams 89
hierarchical organisation of behaviour 12
HTA and part-task training 212
HTA and the design of simulation for
      training 215
Human Error Analysis (HEA) 255
human factors 4
human performance in systems 13
human resource management 5, 153, 232,
      259
hypotheses 18, 22

I

I-A-F model 247
icons 194
information requirements in operations 185
information requirements in plans 193
identifying strategies off line 71, 85
individual differences and variations 69, 140
inferring cognitive operations 71, 72
informant 21
information fragmentation 182
information processing models 29, 247
information requirements 140, 181, 258
inputs and outputs 9, 34
instruction 198, 201
Instructional Systems Development (ISD)
      259
interface design 13, 137, 139, 257

J

job descriptions 141, 237
job design 4, 141, 153, 234
job-aids 5, 142, 153, 223
job-aids and training 226

K

knowledge and constituent skills 208

L

layout of HTA diagrams 89
layout of workplace 140

learning practical skills 198
legal, industrial and cultural compliance 149
liaising with other shifts [task] 133
link analysis 31, 249
lists 97

M

machine packaging operation [task] 34
management [task] 124
managing change [task] 130
manning requirements 234
manuals 223
manufacturing instructions 108, 226
mechanical maintenance [task] 120
metaphors 194
minimal Access Surgery (MAS) [task] 113
minimising distraction 142
minimising error 225
modelling and evaluating strategies 71, 74
monitoring 46, 72, 189
monitoring steady state or rate 189
monitoring to anticipate a target 189

N

neonatal intensive care [task] 83
numbering the HTA 92
nurse in charge of a ward [task] 83, 131
nursing task [task] 124

O

obtaining information 188
on-job instruction 173
Open System Task Analysis 252
operating and design constraints 22
operating constraints 22, 37
operator 21, 245
operator skill 141
operator's goal 22, 32
Organisational Analysis 249, 253

P

P x C rule 27, 37
packaging line [task] 159
part-task training 209, 211, 259

people engaged in the task analysis process
     21
personnel selection 74, 137, 141, 235
planning 192
plans 3, 23, 42
plans in the HTA diagram 90
predictability 81, 147
priming and sharing information 83
procedural cycle plans 52
procedural guides 219
process control tasks 104
public service agencies 179
pure-part training 211

R

railway systems engineer [task] 72
reasons for representing and recording task
     analysis 88
receiving instruction 188
recording information 188
records 222
recoverability 148
redescription 23, 31, 240
redescription has ceased 91
remedial cycle plans 53
representation and feedback 148
representation of information 141
representation of plans 97
representing and situating information and
     control 193
running a shift [task] 132

S

safety critical systems 85
schedule of events 138, 146
sequencing the analysis 93
sequencing training 202
severity of consequences of error 148
SHERPA (Systematic Human Error Reduc-
     tion and Prediction Approach) 255
shift handover [task] 170
shop assistant [task] 200
signal-detection 79, 254
simulation design 213, 259
situating cognition 71, 77

skill-rule-knowledge 248, 254
slow response systems 15, 246
Soft Systems Analysis 253
software to support HTA 255
staff development 172
staff supervision 131
stages in developing a training programme 206
standards 225
stopping redescription 23, 240
stressors 148
subgoals 23
supermarket check-out [task] 47
supervision of a railway system [task] 160, 237
system feedback 14
system goal 21, 32
system life-cycle 137
systems thinking 3, 8, 245

T

tabular Formats for HTA 93
TAG (Task Action Grammar) 249
TAKD (Task Analysis for Knowledge Description) 248
task 17, 22, 246
task analysis 22

task knowledge 248, 249
task system 16, 21
team design 4, 257
team functions 161
time saving 49
timelines 165, 249, 257
training 5, 137, 153, 198, 207, 238
training needs analysis 206
training of occupational therapy students 170
transfer within the task 204

U

ultrasonic inspection [task] 78, 179
underground railway supervision [task] 178
unit operations 105
user 246
using a wordprocessor [task] 117

V

verbal protocol analysis 13, 248

W

withheld information technique 31, 250
workload 138, 164, 257
workplace and interface design 153